Basic Blueprint Reading and Sketching

9th Edition

Thomas P. Olivo

Dr. C. Thomas Olivo

Cengage

Australia • Brazil • Canada • Mexico • Singapore • United Kingdom • United States

Cengage

Basic Blueprint Reading and Sketching,
Ninth Edition
Thomas P. Olivo

Vice President, Career and Professional
   Editorial: Dave Garza

Director of Learning Solutions: Sandy Clark

Senior Acquisitions Editor: Jim DeVoe

Managing Editor: Larry Main

Product Manager: Brooke Wilson, Ohlinger
   Publishing Services

Editorial Assistant: Cris Savino

Vice President, Career and Professional
   Marketing: Jennifer Baker

Marketing Director: Deborah Yarnell

Marketing Coordinator: Mark Pierro

Production Director: Wendy Troeger

Production Manager: Mark Bernard

Content Project Manager: Michael Tubbert

Art Director: Casey Kirchmayer

Technology Project Manager: Joe Pliss

For product information and technology assistance, contact us at
Cengage Customer & Sales Support, 1-800-354-9706
or support.cengage.com.

For permission to use material from this text or product,
submit all requests online at **www.copyright.com.**

Library of Congress Control Number: 2010921508

ISBN-13: 978-1-4354-8378-1
ISBN-10: 1-4354-8378-2

Cengage
200 Pier 4 Boulevard
Boston, MA 02210
USA

Cengage is a leading provider of customized learning solutions
with employees residing in nearly 40 different countries and sales in more
than 125 countries around the world. Find your local representative at:
www.cengage.com.

To learn more about Cengage platforms and services, register or access
your online learning solution, or purchase materials for your course, visit
www.cengage.com.

Printed at CLDPC, USA, 08-22

Basic Blueprint Reading and Sketching

9th Edition

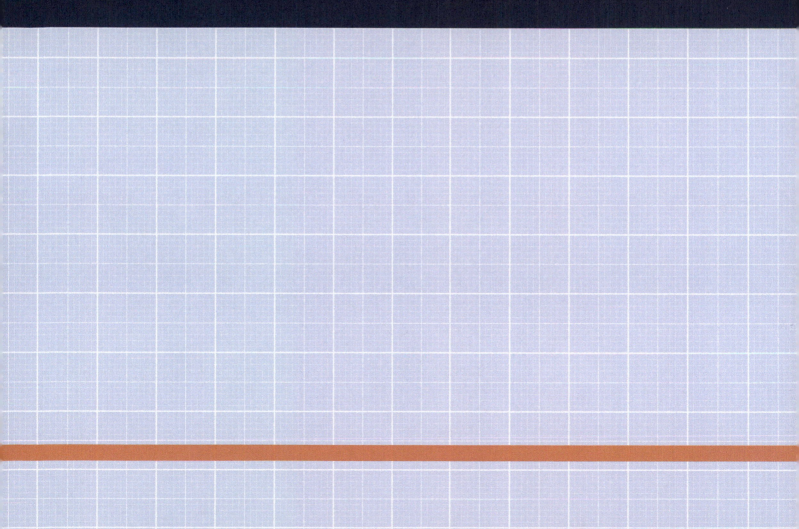

Contents

Preface

Almost all engineers, technicians, architects, machinists, mechanics, and persons whose jobs are related to manufacturing, transportation, construction, and fabrication—from designers to after-market sales persons—frequently refer to blueprints (a copy of any original technical drawing). The written word is insufficient to communicate physical facts completely and accurately. They must be supplemented by graphical representations (drawings). One may never need to make a technical drawing, but all must be able to read and understand them or be professionally illiterate.

This text, *Basic Blueprint Reading and Sketching,* covers all the basic content and exercises to make someone a competent reader of blueprints. It also uniquely covers the increasing need to be able to accurately make technical sketches.

The current edition carries on the traditions that have helped an estimated 58 million users, over its lifetime, to achieve competence in reading and sketching technical drawings as no other book in its field has been able to do. It is the most used blueprint reading book because of the 'hands-on experiences' of the authors and because of Dr. C. Thomas Olivo's original occupational analysis and his writing and organizational skills that are continued by Thomas P. Olivo. In addition, the recommendations by the publisher's reviewers and the author's diverse team of consultants provide continued adjustments to revise the depth and breadth of its content.

ACKNOWLEDGMENTS

For the 9th edition of *Basic Blueprint Reading and Sketching,* special credit goes to my Advisory Committee:

Andrew Howe, Project Manager, BBL Construction Service, Albany, New York

Stephanie Howe, speech language/pathologist, TRAC, Averill Park, New York

Tracy Dunn, educational consultant, Poestenkill, New York

Zachary Dunn, student, Averill Park High School, Averill Park, New York

Jeremy B. Olivo, welder, precision sheet metal operator, machinist, blueprint user at V.M. Choppy & Son, Troy, New York

David Kinerson, electronic equipment designer, NYS DOT, Albany, New York

Todd Reynold, ITT electronics instructor, Albany, New York

Laura Reynolds, editorial services, Rensselaer, New York

Appreciation is expressed to the following industry representatives and their companies for 'going that extra mile' to provide technical resource materials and industrial drawings for several revisions: Eric L. Baker, Dresser/Rand Corporation; Eric Achman, HVAC system designer, T.P. Woodside; Steven Dunn, engineer, Knolls Atomic Research Laboratory; Jennifer Ebanks, owner, Keifito's Plantation Resort, Roatan, Honduras; Joseph D. Harnak, Consulting Engineer, Sperry Rail Service, Inc.; Alan W. Wysocki, The Barden Corporation; Pat Dalton, David Smith, and Gerald Walden, Caterpillar Corporation; Joe Guarino and Robert Dovidio, The Charles Stark Draper Laboratories; Kevin Mangovan, Clausing Industrial, Inc.: Carl Houck, Hamilton Standard Division of United Technologies Corporation; Donald Viscio, Heli-Coil, a Black & Decker Co.; Thomas J. Arconti, Hughes Danbury Optical Systems, Inc; Dan Brookmen, Newport News Shipbuilding Co.; Gary Mohroff, CAD/CAM Instructor, General Electric Company; and Charles Joslin, Fabrication Supervisor Feuz Manufacturing.

Special recognition goes to my chief motivator, Jesus Christ, and to my wife, Stephanie B. Olivo, my mentor, for her editorial/writing services in updating this text; also, to my seven children and nine grandchildren who were able to graciously put up with my time of being 'unavailable' until this revision was completed.

In Memorium

Dr. C. Thomas Olivo

This 9th Edition is dedicated to my father, Dr. C. Thomas Olivo. Without his vision in founding Delmar Publishers, VICA & NOCTI; as president of many vocational education associations; as author and editor of 60 books and 500 publications; and as a consultant to many foreign governments, vocational/occupational education may never have taken hold.

What's New in the Book & Supplements

A perfect depth and breadth of content coverage has been achieved within the 9th edition of *Basic Blueprint Reading and Sketching* by adding three new Part 1 units and a Glossary of selected terms. Although Part 2 has no new sketching units, several new, unique sketching assignments have replaced less appropriate ones.

1. Completely new units are:
   A. Unit 5, *Projection Lines, Other Line Types, and Line Combinations.* Although projection lines are primarily used in sketching, they are frequently used on CAD solid models and 3-D pictorials producing 2-D multi-view drawings.
   B. Unit 18, *Interchangeable Parts, Allowances, and Classes of Fit.* This fills a previous void of how to get from a detail drawing to an assembly drawing. This unit shows how to dimension mating parts not needing GD&T.
   C. Unit 32, *Surface Development and Precision Sheet Metal Drawing.* Following reviewers' feedback, this final missing unit has been added with a representative blueprint and assignment.

2. Six new blueprints have been specifically designed for this revision accompanied with (7) new assignment. The units include: Unit 5, Unit 18, Unit 32, Unit 41, Unit 43, and Unit 46.

3. Section 9 has been renamed *Speciality Drawings.* This now includes *Welding* (Unit 31) and *Surface Development and Precision Sheet Metal Drawing* (Unit 32).

4. Units 29 and 33 have been revised to include more about GD&T.

5. The seven sketching units, 40–46, have been thoroughly revised:
   A. Unit 41, *Oblique Sketching.* This has a new and more applicable assignment.
   B. Unit 43, *Perspective Sketching.* This has a completely new and more appropriate assignment.
   C. Unit 46, *Proportions and Assembly Drawings.* This has a new assembly assignment.

6. All figures and drawings have been checked and changed in places, enhancing clarity to more clearly represent the best drafting techniques. Over 180 drawings and figures have been modified to achieve this objective.

7. Numerous drawings throughout this book have been revised to conform to ANSI Y14-M standards and ISO metric standards.

8. All GD&T related content, charts, and drawings in Unit 29 has been extensively updated to reflect the latest ANSI standards.

9. All references to computers used in drawing, storage, CNC & MFG, CAD/CAM, CADD, CIM, robotics, scanning, and copying have been updated and revised.

10. *Tonal Shading* has been added to 17 blueprints, enhancing visualizations and comprehension.

11. New text layout and color in text, blueprints, and assignments should make reading much easier.

12. A new e.resource CD supplement has been developed.

13. Accurate answers to all assignments has been created by the author.

Guidelines for Study

Basic *Blueprint Reading and Sketching* is designed to be a stand-alone text for developing skills in reading and accurately interpreting industrial drawings for manufacturing, construction, and transportation curricula for high school students through college level. It is also used for preparing simple technical sketches and has been used as a supplement in several program areas, such as CAD/CIM/CADD, drafting, robotics, welding, machining, plumbing, carpentry, electrical, stage design, other construction courses, and numerous transportation courses. Another practical application of this text is for use in training programs for a variety of courses given by private trade schools, manpower training programs, military, and government agencies. By integrating the content of this blueprint and sketching book with other technical drawing courses, the ability of the student to prepare many types of specialized drawings and working sketches is enhanced and accelerated.

The use of a unit pattern of organization permits the contents to be readily adaptable to functional courses within educational, business and industry, government, and other special training programs. The text-workbook is widely used in class, group, and individualized/self-paced instruction. However, this is not a "do-it-yourself" text; it requires a knowledgeable instructor or mentor.

Students may also start the course with technical sketching and finish with blueprint reading, since both parts of the book progress from a simple part and end with a multi-part assembly assignment.

ORGANIZATION OF THE TEXTBOOK

Basic Blueprint Reading and Sketching contains two major parts. **Part 1** covers basic principles of blueprint reading, with companion drawings and assignments relating to applications of each new principle. **Part 2** deals with principles and techniques of preparing technical sketches, without the use of instruments.

The instructional units in **Part 1** and **Part 2** are arranged in a natural sequence of teaching/learning difficulty. Each unit includes **basic principles,** a **blueprint,** and an **assignment** as listed in the Contents.

UNIT BASIC PRINCIPLES

Blueprint reading and sketching principles, concepts, ANSI/ASMF and ISO Metric drafting symbols and standards, terminology, manufacturing process notes, and other related technical information contained on a mechanical or CAD-produced engineering drawing are all described in detail in the basic principles.

UNIT BLUEPRINT(S)

Each new principle is applied on one or more industrial drawings. Many of the drawings in the beginning units are simplified by removing the Title Block. The representative drawings appear in the **blueprint(s).** While the drawings are based on ANSI/ASMF and ISO Metric measurement systems and representation practices, there are adaptations of the standards as found in industry. CAD pictorials of finished parts are now provided on many of the **blueprint(s)**. The pictorials provide a learning tool by identifying the shape, features, dimensions, and other characteristics of each part.

An image library on the e.resource CD provides a teaching tool for increasing instructional efficiency and accelerating the student's mastery in reading and interpreting each blueprint.

UNIT ASSIGNMENT(S)

Problems and other test items are provided in the unit assignment, covering all basic principles that are applied in the blueprints. The assignments are stated in the terminology of the workplace and deal with principles, practices, and the kinds of experiences needed to read drawings for dimensions, shape descriptions, machining operations, and other essential data as required of technicians, craftpersons, related skilled workers, and professionals.

Use is made of encircled letters, numbers, leaders (**callouts**), on each drawing in the unit blueprint(s) and assignment(s). The callouts simplify the statement of problems. A ruled space is included with the unit assignment for ease in recording each answer.

The sketching assignments deal with principles and applications of sketching, lettering, and techniques used in making orthographic, oblique, isometric, perspective, pictorial, and assembly sketches.

E.RESOURCE CD SUPPLEMENT

This is an educational resource that creates a truly electronic classroom. It is a CD-ROM containing tools and instructional resources that enrich the classroom and make the instructor's preparation time shorter. The elements of *e.resource* link directly to the text and tie together to provide a unified instructional system. With this *e.resource,* instructors can spend their time teaching, not preparing to teach. ISBN: 1-4018-4880-X.

Features contained in *e.resource* include:

- **Syllabus:** Lesson plans are created by unit, giving the option of using these lesson plans with specialized course information.

- **Unit Hints:** Objectives and teaching hints provide the basis for a lecture outline that helps present concepts and material. Key points and concepts can be graphically highlighted for student retention.

- **Instructor's Manual:** This PDF file contains answers and solutions to the unit assignments.

- **PowerPoint® Presentation:** These slides provide the basis for a lecture outline that helps present concepts and material. Key points and concepts can be graphically highlighted for student retention.

- **Optical Image Library:** This is a database of key images taken from the text that can be used in lecture presentations, tests and quizzes, and to enhance PowerPoint presentations.

- **Exam View Test Bank:** Over 500 questions of varying levels of difficulty are provided in true/false, multiple choice, fill-in-the-blank, and short answer formats to assess student comprehension. This versatile tool enables the instructor to manipulate the data to create original tests.

Bases for Blueprint Reading and Sketching

Technical information about the shape and construction of a simple part may be conveyed from one person to another by the spoken or written word. As the addition of details makes the part or mechanism more complex, the designer, draftsperson, engineer, technician, and mechanic must use precise methods to describe the object adequately.

Although a picture drawing (Figure 1–1) or a photograph may be used to describe a part, neither method of representing an object shows the exact sizes, interior details, construction, or machining operations that are required. On the other hand, a drawing made accurately with instruments (Figure 1–2) or produced by computer or a technical sketch meets the requirements for an accurate description of shape, construction, size, and other details.

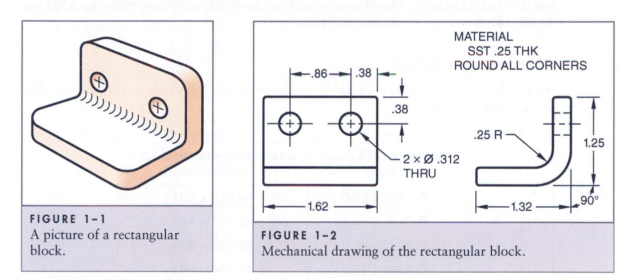

FIGURE 1–1
A picture of a rectangular block.

FIGURE 1–2
Mechanical drawing of the rectangular block.

BLUEPRINT READING AS A UNIVERSAL LANGUAGE

Blueprints provide a universal language by which all information about an object is furnished to the craftspeople, technicians, designers, assemblers, machine operators, and programmers, as well as after-market repairs and maintenance persons. Blueprint reading refers to the process of interpreting a drawing. An accurate mental picture of how the object will look when completed can be formed from the information presented.

Training in blueprint reading includes developing the ability to visualize various manufacturing and fabricating processes required to make a part and to apply the basic principles of drafting that underlie the use of different lines, surfaces, and views. The training also includes how to apply dimensions, take measurements, and visualize how the inside of a part or mechanism looks. An understanding of universal measurements and other standards, symbols, signs, and techniques the draftsperson/designer uses to describe a part, unit, or mechanism completely must be developed. Technicians also develop fundamental skills in making sketches so that data relating to dimensions, notes, and other details needed to construct or assemble a part can be recorded on a sketch.

STANDARDS GOVERNING THE PREPARATION OF DRAWINGS

Because all industrialized nations prepare technical drawings according to universally adopted standards, symbols, technical data, and principles of graphic representation, blueprints may be uniformly interpreted throughout the world. This fact translates to mean that parts, structures, machines, and all other products (that are designed according to the same system of measurement) may be accurately manufactured and are interchangeable.

Standards-Setting Organization

The American National Standards Institute (ANSI) and the International Organization for Standardization (ISO) have adopted drafting standards that are voluntarily accepted and widely used throughout the world. These systems incorporate and complement other engineering standards that are developed and accepted by such professional organizations as the American Society of Mechanical Engineers (ASME), the American Welding Society (AWS), the American Institute for Architects (AIA), and others. These standards-setting bodies deal with specific branches of engineering, science, and technology. Some large companies have adopted their own standards to suit their individual needs. Many blueprints and sketches in this text closely follow the ANSI and ISO standards and the current industrial practices.

Standards for Sheets

Standards have been published by ASME for the sheets onto which blueprints are made (Figure 1–3 and Figure 1–4).

Standard USA Size* (inch)	Nearest International Size** (millimeter)
A 8.5 × 11.0	A4 210 × 297
B 11.0 × 17.0	A3 297 × 420
C 17.0 × 22.0	A2 420 × 594
D 22.0 × 34.0	A1 594 × 841
E 34.0 × 44.0	A0 841 × 1189

*ASME Y14.1, **ASME Y14.1M

FIGURE 1–3

COMMON ELEMENTS IN DRAFTING

Close examination of a technical drawing reveals that a part or an assembly of many parts must be clearly and accurately represented by applying principles, techniques, and other common elements of drafting, as follows:

■ Lines that, when combined with each other, provide graphic information about the external shape and details of an object.

■ Dimensions that give the size and location of parts, shapes, and assemblies.

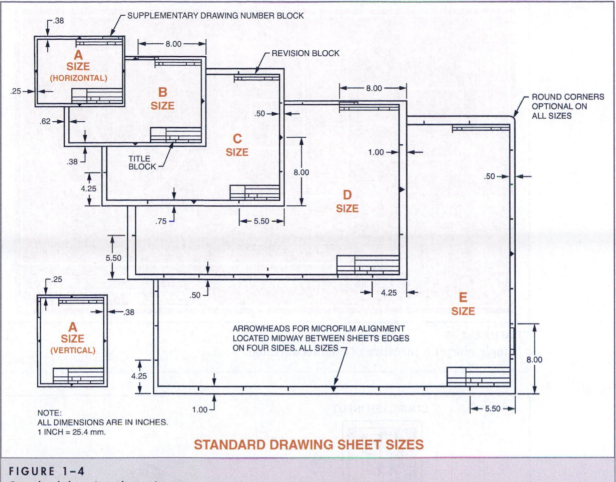

STANDARD DRAWING SHEET SIZES

FIGURE 1-4
Standard drawing sheet sizes.

- **Sections** that are used to expose internal construction details when an object is cut apart by imaginary cutting planes.
- **Processes** that are represented by symbols and notes for purposes of identifying how each part is to be produced.
- **Geometric positioning** that deals with placing and accuracy of geometric objects in relationship to one another.
- **Notes and other non-graphical information** clearly lettered (printed).
- **Tolerances** that deal with accuracy related to form and size and position of finished surfaces, where high precision is required.
- **Materials** that are used for making parts, including information about weight, strength, and hardness.
- **Computer-aided drafting** (CAD) applications, where appropriate.
- **Technical sketching** techniques for preparing line and form engineering and technical sketches to complement precise engineering drawings.

INDUSTRIAL PRACTICES RELATING TO DRAWINGS

Drawing Reproduction Processes

Traditionally, a blueprint of an original drawing, which may be made on high-quality paper, vellum, and acetate or polyester film, is produced by exposing the drawing to a strong light source. Some of

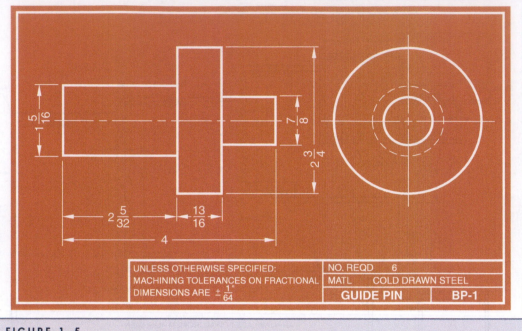

FIGURE 1–5
Sample blueprint (negative of original drawing).

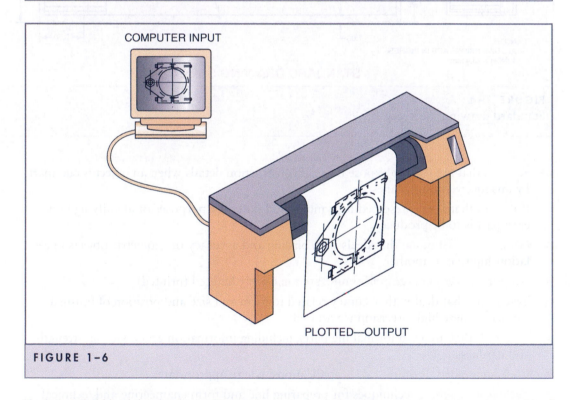

FIGURE 1–6

the light is held back by the drawn lines as it passes through a transparent tracing onto a sensitized blueprint paper. After exposure, the blueprint paper is passed through a developer fluid, washed by water, moved through a fixative (to set the color), rewashed, and dried. This blueprint (except for slight shrinkage during the drying stage) is an exact duplicate but in negative form (Figure 1–5).

The term blueprint originally referred to a blue and white reproduction of a drawing. Today, it is used loosely in modern industrial language to include any type of reproduction of a technical drawing, including computer-created ones. (Figure 1–6).

The highest quality reproductions are made by plotters of various types: the flatbed plotter that holds the paper (usually A-size or B-size) on a flat surface while a pen moves in two directions; a drum plotter where the paper is attached to a rotating drum producing horizontal ("X" axis) lines as a pen moves across the drum to reproduce vertical ("Y" axis) lines; or a multi-grip plotter that uses small rubber rollers to grip the paper at the edges, as opposed to having the paper attached to the circumference of the drum. Simultaneous moving of the drum and the pen on both of these produces curves and angle lines. The size of the paper used in the drum plotter is limited by the width and the circumference of the drum, which can be designed to use up through E-size or AO drawings.

Today, most plotters operate by commands from a computer. They can hold any number of pens up to 20 and produce very accurate continuous line drawings with varying lineweight and color.

Printers are also used to reproduce drawings. They are generally less expensive than plotters but are not as accurate because they can only simulate lines by printing a closely spaced series of dots.

Another method of reproduction known as diazo continues to be popular. This process projects light through the original drawing, which is in direct contact with the recipient paper or transparency. That paper is then developed with an ammonia process to show the image, usually in black, brown, or blue. A variety of paper weights and finishes as well as transparencies can be used in diazo.

Copying machines, similar to ones used for typical office copying, have also been designed to handle blueprints and engineering prints of varying sizes. There are several advantages to these machines—they are relatively inexpensive, they do not require coated copy materials or chemicals, and the size of the drawing can be enlarged or reduced as desired.

Another style of copy machine is also used. It uses a wide-bed page scanner coupled with a wide format printer. In this type of copy system, resolution is limited to that of the scanner, typically 300 dots per inch. Typical wide-bed printer resolution is around 1200 dots per inch. This means that the smallest line the printer can create is 0.00083.

Drawing Storage Procedures

Original manually created drawings are seldom sent to the shop, job site, or laboratory. Instead, exact reproductions (called blueprints or whiteprints) are made of the original. Duplicate copies are usually distributed among all individuals who are involved in planning, manufacturing, assembling, or other work with a part or unit. The original drawings are filed for record purposes and for protection.

The options available for storage of drawings and blueprints have increased as technology has advanced. Microfilm has been one of the most widely used procedures for many years and is still used in some places today. In this method, a greatly reduced photographic reproduction is made on film and then mounted into cardboard aperture cards. These cards are easy to file and keep track of. Many feel that they are also more durable.

Blueprints are also saved in special storage devices, including flat-drawer files, files designed for vertical hanging, or tubes. Drawings stored in these ways must also be available for rapid and convenient retrieval; therefore, prints are often folded and stored in standard office file cabinets. If properly organized, this method does provide rapid retrieval and security.

For many companies, the preferred storage method is electronic, whether the drawing is produced by hand or by CAD (Computer Aided Drafting). With CAD, the drawing can be saved directly to a disk or it can be produced as hardcopy (blueprint) by one of several types of printers. Scanners of various types can be used to convert handmade drawings into computer files for disk storage.

Computer files can be stored on a number of different types of media. Common electronic storage media include USB flash drives, 3.5" floppy disks, hard drives, optical disks, and magnetic tape. Generally, when a drawing is being edited, it is saved frequently to a hard drive located on a network server.

A large company will have two network servers: a primary server and a backup server. Companies typically prefer documents to be saved on the server, not on the user's computer hard drive. Hard drive failure on individual computers is common and costly, so companies store the information on the servers.

With the use of the Internet, a draftsperson can convert a CAD file to an Adobe PDF file, which can be emailed to a recipient so the drawing can be viewed on a remote computer. The drawing can be printed on a large format printer. Adjustments can be made to the drawing by sketching the comments. The drawing can then be scanned back into the computer and be sent back to the original draftsperson. The comments could then be taken into account in the revised CAD file.

Some consider electronic storage to be superior to former methods because it reduces deterioration and the possibility of loss. Also, when drawings are saved as hard copies (prints), there is inevitable wear and tear on them, which compromises the quality of any reproduction. Drawings saved electronically can be consistently reproduced with the same quality. It is also possible to revise and correct drawings which have been saved on magnetic media (magnetic tape, hard drives). However, disks and computer systems can develop problems and the longevity of magnetic media or optical disks has not been proven; therefore, there continues to be a need for a choice.

Hardcopy and Softcopy Drawings

The term blueprint or print identifies drawings that may be produced by drafting room techniques or are computer-generated. Blueprint reading traditionally involves the physical handling of a sheet on which a drawing is reproduced, but it also refers to drawings as they appear on computer screens. The latter is called softcopy. The drawing sheet is sometimes referred to as hardcopy and is a permanent copy of a drawing. Hardcopy may be produced on conventional blueprinting equipment or it may be computer-generated. Drawings in hardcopy form may be reproduced in one color or an assortment of colors.

CAD/CAM

Some industries tie their engineering and manufacturing departments together using a CAD/CAM linkup over a computer network. CAM stands for computer-aided manufacturing. In this process, the drafting department designs a product on a computer. After completion, the drawing is converted to coordinate data and then stored. Additional software called a post processor then converts the stored coordinate data into a usable format for the machines in the manufacturing department. Hardcopy is not produced unless someone needs to study the drawing away from a computer terminal.

SOME TECHNICAL INFORMATION SUPPLIED ON DRAWINGS

The positive form blueprint of the guide pin (Figure 1–7) illustrates other information that generally appears on a drawing. A title block is included that contains the following. (See pages 00 for a more detailed title block.)

- Name of the part (guide pin)
- Quantity needed (6)
- Drawing number (BP–1)
- Dimensional tolerance (± 1/64")
- Material (cold drawn steel)

UNIVERSAL SYSTEMS OF MEASUREMENT

Drawings contain dimensions that identify measurements of straight and curved lines, surfaces, areas, angles, and solids. Linear or straight line measurements between two points, lines, or surfaces are most widely used. Every linear measurement begins at a particular reference point and ends at a measured point. These references identify particular features as illustrated in Figure 1–8. Characteristics of linear measurements are summarized in the second drawing at (B).

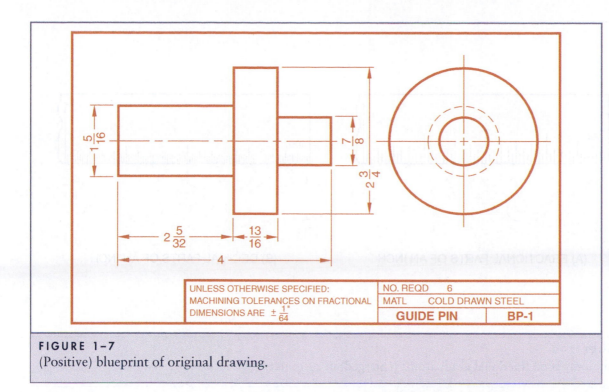

FIGURE 1-7
(Positive) blueprint of original drawing.

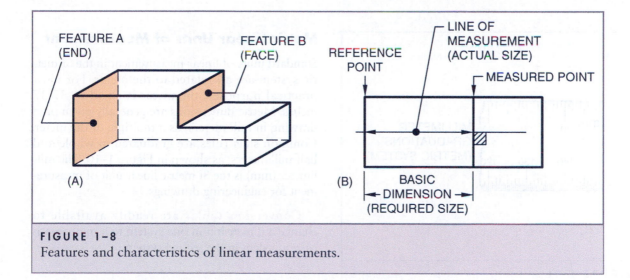

FIGURE 1-8
Features and characteristics of linear measurements.

The straight line distance between these two points is the **line of measurement.** It represents the actual required dimension and is referred to as the **basic size.** A basic size is sometimes represented on a drawing by this ⬜ symbol. The basic size is enclosed within the symbol as, for example, 1.625 . The degree of accuracy to which a part is to be machined, fitted, formed, or assembled relates to its basic size.

Basic Units of Linear Measurement

Dimensions appear on drawings in one or a combination of two different units of measurement, as in the case where inch and metric units are used together. Customary inch (British–United States standard) dimensions are given as fractional parts of an inch or as decimal inch values. The inch is subdivided into equal parts, called **fractional parts or decimal (mils) parts.** Figure 1-9A illustrates commonly used fractional parts of an inch (1/64″, 1/32″, 1/16″, 1/8″, 1/4″, and 1/2″). The graduations on the steel rule in Figure 1–9B show decimal parts of an inch (.010″ [ten mils], .050″ [fifty mils], .100″ [100 mils], and .500″ [500 mils].

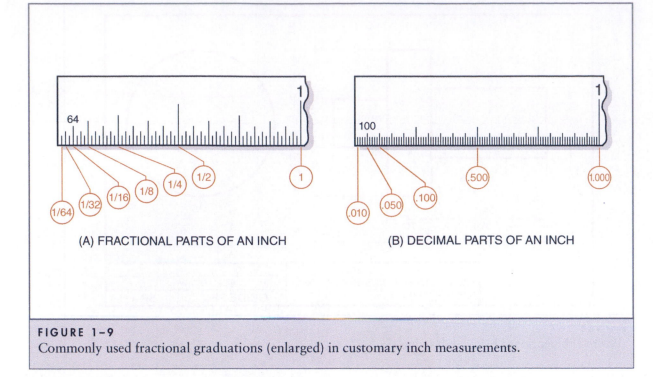

(A) FRACTIONAL PARTS OF AN INCH

(B) DECIMAL PARTS OF AN INCH

FIGURE 1-9
Commonly used fractional graduations (enlarged) in customary inch measurements.

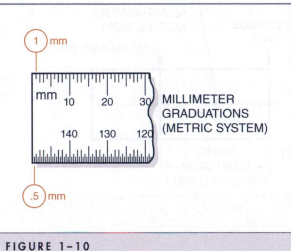

MILLIMETER GRADUATIONS (METRIC SYSTEM)

FIGURE 1-10
Whole and half millimeter (mm) graduations (enlarged).

Metric Linear Units of Measurement

Standard units of linear measurement in the **SI** metric system are all related to the **meter.** For most practical purposes, the meter is equal to 39.37 inches. Metric dimensions are generally given on a drawing in millimeters, as a multiple of the meter. Common steel rules are graduated in whole and half millimeters, as shown in Figure 1–10. The millimeter (mm) is the SI metric linear unit of measurement for engineering drawings.

Conversion tables are readily available to change a dimension in one system to a mathematical equivalent in the second system.

ASSIGNMENT—UNIT 1: BASES FOR BLUEPRINT READING AND SKETCHING

Student's Name _____

1. Name one difference between picture drawings or photographs and a technical drawing of an object.

   1. _____
      _____

2. Tell briefly why blueprints provide a universal language.

   2. _____
      _____

3. Name two organizations that develop drafting standards.

   3. (a) _____
      (b) _____

4. List three common elements of drafting that are used to produce technical drawings.

   4. (a) _____
      (b) _____
      (c) _____

5. State one reason why printers are not as accurate as plotters for reproducing technical drawings.

   5. _____
      _____

6. Name two reproduction methods other than plotters and printers.

   6. (a) _____
      (b) _____

7. Provide two advantages of disk storage over storage of original drawings on paper.

   7. (a) _____
      (b) _____

8. Explain the difference between hardcopy and softcopy.

   8. _____
      _____

9. Name the end points to every linear measurement.

   9. _____ _____

10. Describe what the term "line of measurement" means.

    10. _____
        _____

11. Tell what function a basic size dimension serves.

    11. _____
        _____

12. Briefly explain how an off-site computer can be used to modify a drawing held at another computer location.

    12. _____
        _____
        _____

13. Name the two major linear measurement systems.

    13. (a) _____
        (b) _____

14. Give an example of a fractional part of the basic unit of measurement in each system.

    14. (a) _____
        (b) _____

15415-1   T013   02

Blueprint
Reading

Lines

The Alphabet of Lines and Object Lines

The line is the basis of all industrial drawings. By combining lines of different thicknesses, types, and lengths, it is possible to describe graphically any object in sufficient detail so that people with a basic understanding of blueprint reading can accurately visualize the shape of the part.

THE ALPHABET OF LINES

The American National Standards Institute (ANSI) has adopted and recommended certain drafting techniques and standards for lines. The types of lines commonly found on drawings are known as the alphabet of lines. The six types of lines that are most widely used from this alphabet include: (1) object lines, (2) hidden lines, (3) center lines, (4) extension lines, (5) dimension lines, and (6) projection lines. A brief description and examples of each of the six types of lines are given in this section. These lines are used in combination with each other on all the prints in the Blueprint Series. Problem material on the identification of lines is included in the Assignment Series. Other lines such as those used for showing internal part details or specifying materials are covered in later units.

The thickness (weights) of lines are relative because they depend largely on the size of the drawing and the complexity of each member of an object. For this reason, comparative thicknesses (weights) of lines are used. ANSI recommends two line thicknesses: thick and thin. Thick lines are two times the thickness of thin. Drawings used to meet military standards have three line thicknesses: thin, medium, and thick.

OBJECT LINES

The shape of an object is described on a drawing by thick (dark) lines known as **visible edge** or
object lines. An object line, Figure 2–1, is always drawn thick (dark) and solid so that the outline
or shape of the object is clearly emphasized on the drawing, Figure 2–2.

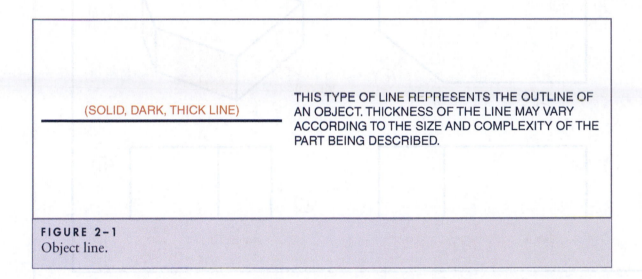

(SOLID, DARK, THICK LINE)

THIS TYPE OF LINE REPRESENTS THE OUTLINE OF
AN OBJECT. THICKNESS OF THE LINE MAY VARY
ACCORDING TO THE SIZE AND COMPLEXITY OF THE
PART BEING DESCRIBED.

FIGURE 2–1
Object line.

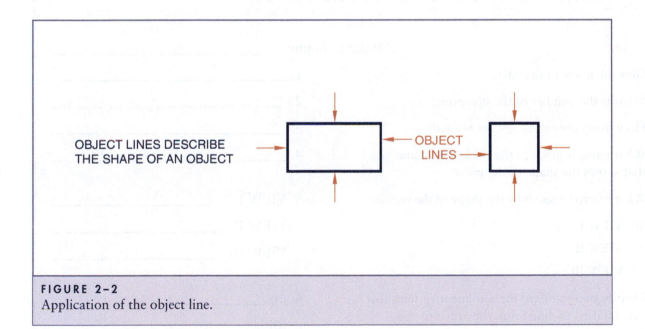

OBJECT LINES DESCRIBE
THE SHAPE OF AN OBJECT

OBJECT
LINES

FIGURE 2–2
Application of the object line.

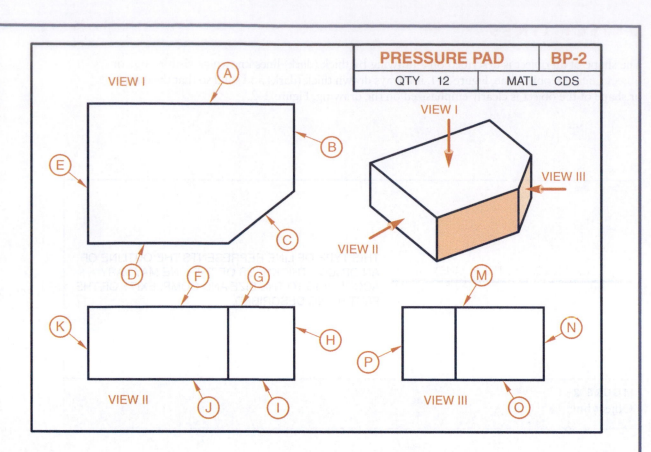

PRESSURE PAD	BP-2
QTY 12	MATL CDS

ASSIGNMENT—UNIT 2: PRESSURE PAD (BP-2)

Student's Name _____

1. Give the name of the part.   1. _____

2. What is the number of the blueprint?   2. _____

3. How many **pressure pads** are needed?   3. _____

4. What name is given to the thick (dark) line   4. _____
   that shows the shape of the part?

5. What lettered lines show the shape of the part in: 5. VIEW I   _____

   (a) VIEW I   VIEW II _____

   (b) VIEW II   VIEW III _____

   (c) VIEW III

6. Identify two standard measurement systems that   6. (a) _____
   may be used to dimension the **pressure pad**
   drawing.   (b) _____

7. Name two basic processes for producing multiple   7. (a) _____
   copies of a finished drawing.
                                                            (b) _____

Hidden Lines and Center Lines

HIDDEN LINES

To be complete, a drawing must include lines that represent all the edges and intersections of surfaces in the object. Many of these lines are invisible to the observer because they are covered by other portions of the object. To show that a line is hidden, the draftsperson usually uses a thin, broken line of short dashes, Figure 3–1. Figure 3–2 illustrates the use of hidden lines.

(THIN, DARK, BROKEN LINE, SHORT DASHES)   THIS TYPE OF LINE REPRESENTS
INVISIBLE EDGES AND SURFACES

FIGURE 3–1
Hidden line.

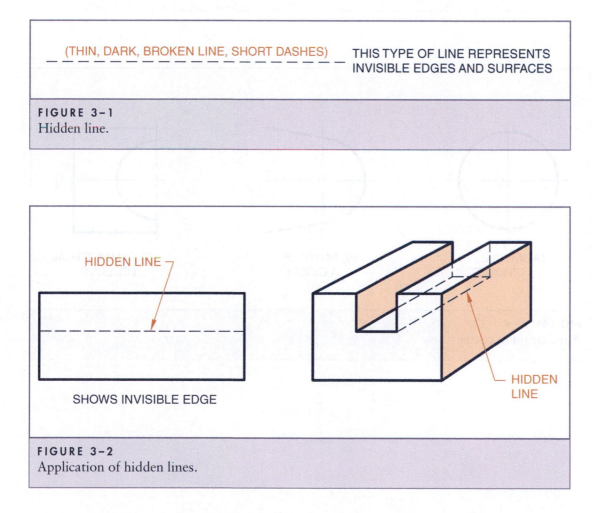

HIDDEN LINE

SHOWS INVISIBLE EDGE

HIDDEN LINE

FIGURE 3–2
Application of hidden lines.

CENTER LINES

A center line, Figure 3–3, is drawn as a thin (light), broken line of long and short dashes, spaced alternately. Center lines are used to indicate the center of a whole circle or a part of a circle and also to show that an object is symmetrical about a line, Figure 3–4. The symbol ℄ is often used with a center line.

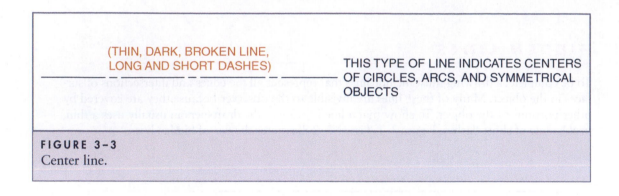

(THIN, DARK, BROKEN LINE, LONG AND SHORT DASHES)

THIS TYPE OF LINE INDICATES CENTERS OF CIRCLES, ARCS, AND SYMMETRICAL OBJECTS

FIGURE 3–3
Center line.

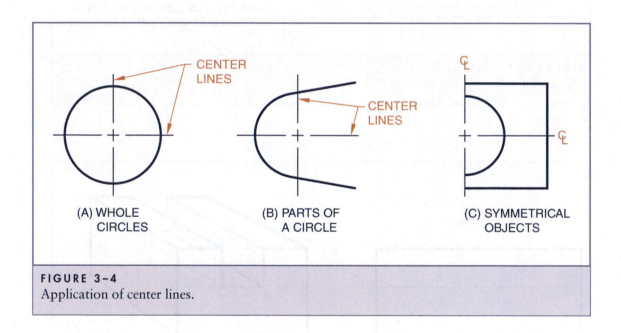

CENTER LINES

CENTER LINES

(A) WHOLE CIRCLES

(B) PARTS OF A CIRCLE

(C) SYMMETRICAL OBJECTS

FIGURE 3–4
Application of center lines.

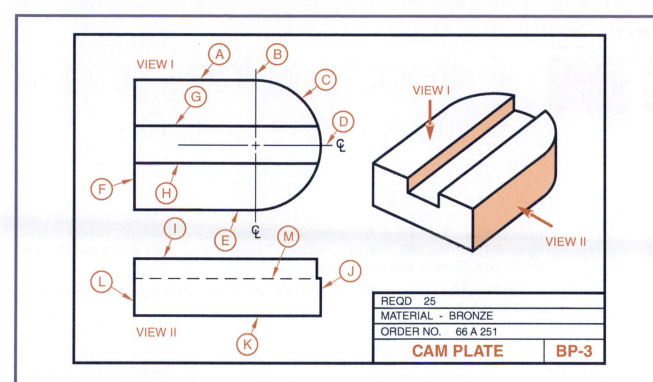

REQD   25
MATERIAL - BRONZE
ORDER NO.   66 A 251

CAM PLATE   **BP-3**

ASSIGNMENT — UNIT 3: CAM PLATE (BP-3)

Student's Name _____

1. How many cam plates are required?   1. _____

2. Name the material for the parts.   2. _____

3. What type of line is used to describe the shape of the part?   3. _____

4. Give the letters of all the lines that show the outside shape of   4. _____
   the part.
                                                                            _____

5. Name the kind of line that represents an invisible edge.   5. _____

6. What lettered lines in VIEW I show the slot in the cam plate?   6. _____

7. What line in VIEW II shows an invisible surface?   7. _____

8. What kind of line indicates the center of a circle or part of a circle? 8. _____

9. What lettered line in VIEW I shows the circular end?   9. _____

10. What lettered lines in VIEW I are used to locate the center of the   10. _____
    circular end?

Extension Lines and Dimension Lines

EXTENSION LINES

Extension lines are used in dimensioning to show the size of an object. Extension lines, Figure 4–1, are thin, dark, solid lines that extend away from an object at the exact places between which dimensions are to be placed.

A space of 1/16 inch is usually allowed between the object and the beginning of the extension line, Figure 4–2.

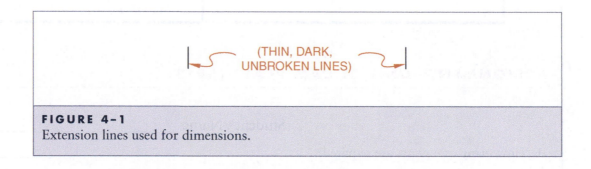

FIGURE 4–1
Extension lines used for dimensions.

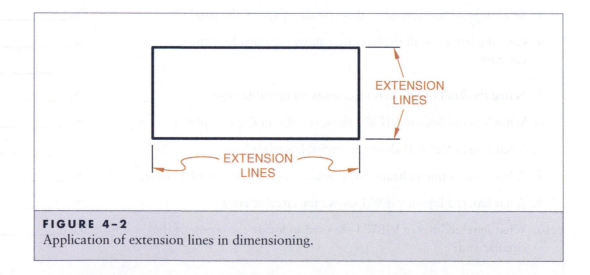

FIGURE 4–2
Application of extension lines in dimensioning.

DIMENSION LINES

Once the external shape and internal features of a part are represented by a combination of lines, further information is provided by dimensions. Fractional, decimal, and metric dimensions are used on drawings to give size descriptions. Each of these three systems of dimensioning is used throughout this text.

Dimension lines, Figure 4–3, are thin lines broken at the dimension and ending with arrowheads. The tips or points of these arrowheads indicate the exact distance referred to by a dimension placed at a break in the line.

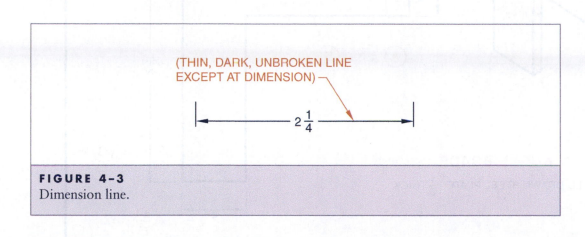

(THIN, DARK, UNBROKEN LINE EXCEPT AT DIMENSION)

$2\frac{1}{4}$

FIGURE 4–3
Dimension line.

The point or tip of the arrowhead touches the extension line. The size of the arrow is determined by the thickness of the dimension line and the size of the drawing. Closed (←) and open (→) arrowheads are the two shapes generally used. The closed arrowhead is preferred. The extension line usually projects 1/16 inch beyond a dimension line, Figure 4–4. Any additional length to the extension line is of no value in dimensioning.

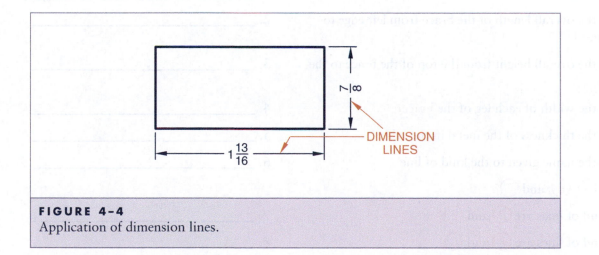

$\frac{7}{8}$

$1\frac{13}{16}$

DIMENSION LINES

FIGURE 4–4
Application of dimension lines.

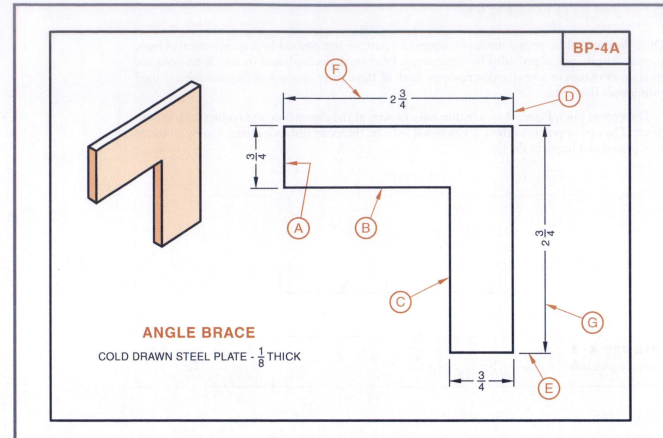

ANGLE BRACE

COLD DRAWN STEEL PLATE - $\frac{1}{8}$ THICK

BP-4A

ASSIGNMENT A—UNIT 4: ANGLE BRACE (BP-4A)

Student's Name _____

1. Name the material specified for the angle brace.

   1. _____

2. What is the overall length of the brace from left edge to right edge?

   2. _____

3. What is the overall height from the top of the brace to the bottom?

   3. _____

4. What is the width of each leg of the brace?

   4. _____

5. What is the thickness of the metal in the brace?

   5. _____

6. What is the name given to the kind of line marked Ⓐ, Ⓑ, and Ⓒ?

   6. _____

7. What kind of lines are Ⓓ and Ⓔ?

   7. _____

8. What kind of lines are Ⓕ and Ⓖ?

   8. _____

9. Why are object lines made thicker than extension and dimension lines?

   9. _____

   _____

   _____

   _____

   _____

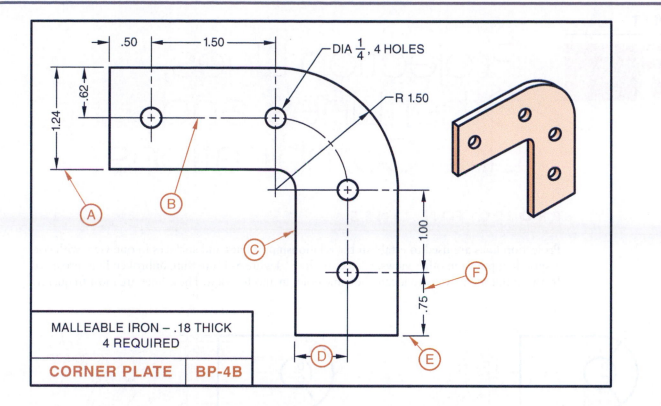

MALLEABLE IRON – .18 THICK
4 REQUIRED

CORNER PLATE | **BP-4B**

ASSIGNMENT B—UNIT 4: CORNER PLATE (BP-4B)

Student's Name _____

1. What kind of line is Ⓐ?   1. _____

2. What kind of line is Ⓑ?   2. _____

3. What kind of line is Ⓒ?   3. _____

4. Determine the overall length of the plate from left edge to   4. _____
   right edge. Note: R = Radius.

5. Determine the overall height of the plate from top to bottom.   5. _____

6. Give the center distance between the two upper holes.   6. _____

7. Determine distance Ⓓ.   7. _____

8. What kind of line is Ⓔ?   8. _____

9. What kind of line is Ⓕ?   9. _____

10. What radius forms the rounded corner of the plate.   10. _____

11. Name the material specified for the corner plate.   11. _____

12. How many corner plates are required?   12. _____

5

Projection Lines, Other Lines, and Line Combinations

PROJECTION LINES

Projection lines are used to establish the relationship of lines and surfaces in one view with corresponding points in other views. **Projection lines,** Figure 5–1, are thin, unbroken lines projected from a point in one view to locate the same point in another view. These lines are most frequently

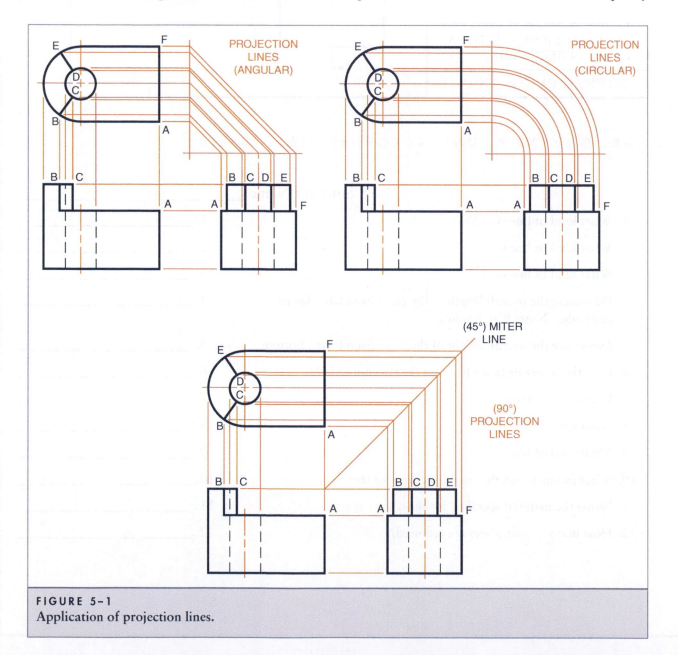

FIGURE 5–1
Application of projection lines.

used when making sketches and in pattern development drawings. Projection lines do not appear on finished drawings except where a part is complicated and it becomes necessary to show how certain details on a drawing are obtained.

OTHER LINES

In addition to the six common types of lines, the alphabet of lines also includes other types, such as the cutting plane and viewing plane lines, break lines, and phantom lines. These less frequently used lines, Figure 5–2, are found in more advanced drawings and will be described in greater detail as they are used in later drawings in this text.

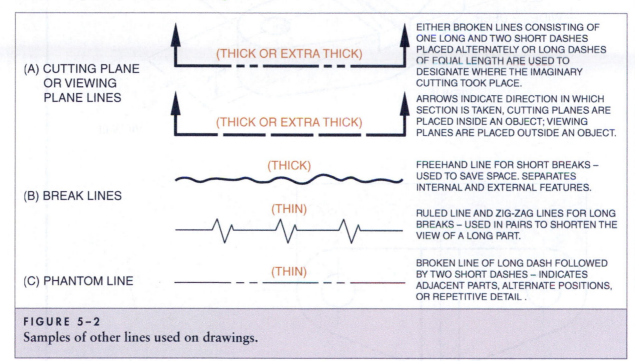

(A) CUTTING PLANE OR VIEWING PLANE LINES

(THICK OR EXTRA THICK)

EITHER BROKEN LINES CONSISTING OF ONE LONG AND TWO SHORT DASHES PLACED ALTERNATELY OR LONG DASHES OF EQUAL LENGTH ARE USED TO DESIGNATE WHERE THE IMAGINARY CUTTING TOOK PLACE.

(THICK OR EXTRA THICK)

ARROWS INDICATE DIRECTION IN WHICH SECTION IS TAKEN, CUTTING PLANES ARE PLACED INSIDE AN OBJECT; VIEWING PLANES ARE PLACED OUTSIDE AN OBJECT.

(B) BREAK LINES

(THICK)

FREEHAND LINE FOR SHORT BREAKS – USED TO SAVE SPACE. SEPARATES INTERNAL AND EXTERNAL FEATURES.

(THIN)

RULED LINE AND ZIG-ZAG LINES FOR LONG BREAKS – USED IN PAIRS TO SHORTEN THE VIEW OF A LONG PART.

(C) PHANTOM LINE

(THIN)

BROKEN LINE OF LONG DASH FOLLOWED BY TWO SHORT DASHES – INDICATES ADJACENT PARTS, ALTERNATE POSITIONS, OR REPETITIVE DETAIL .

FIGURE 5–2
Samples of other lines used on drawings.

LINES USED IN COMBINATION

Most drawings consist of a series of object lines, hidden lines, center lines, extension lines, and dimension lines used in combination with each other to give a full description of a part or mechanism, Figure 5–3.

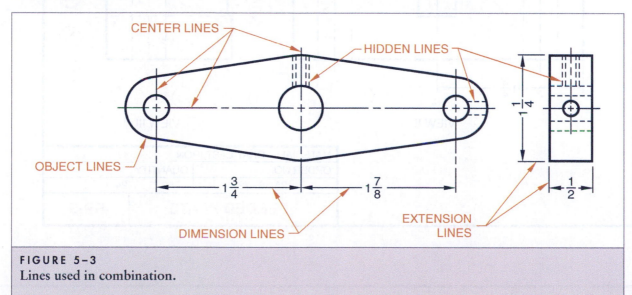

CENTER LINES

HIDDEN LINES

OBJECT LINES

$1\frac{3}{4}$

$1\frac{7}{8}$

$1\frac{1}{4}$

$\frac{1}{2}$

DIMENSION LINES

EXTENSION LINES

FIGURE 5–3
Lines used in combination.

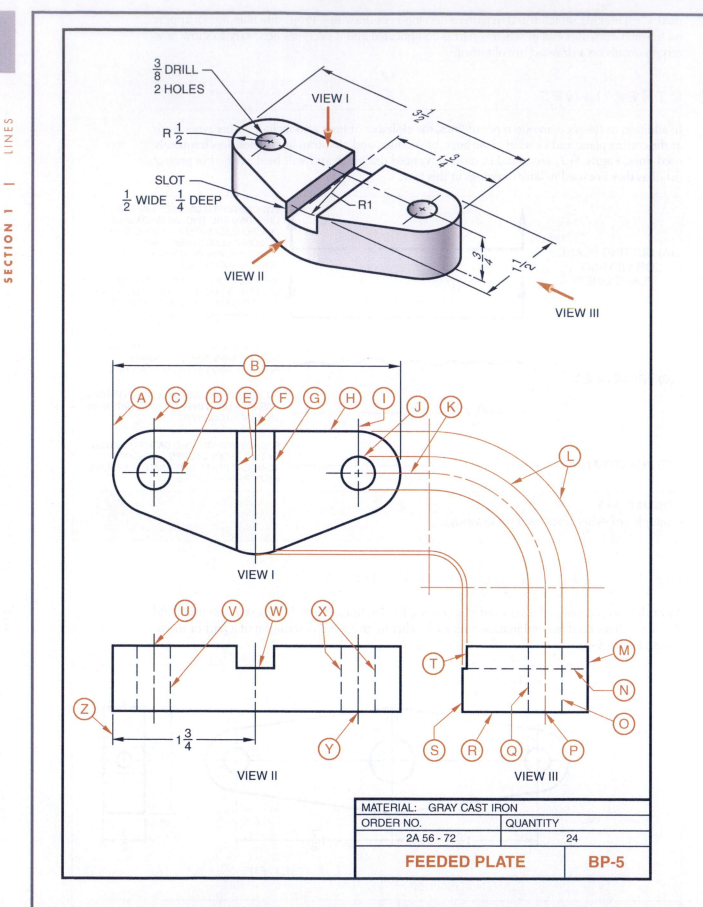

$\frac{3}{8}$ DRILL
2 HOLES

R $\frac{1}{2}$

VIEW I

SLOT
$\frac{1}{2}$ WIDE $\frac{1}{4}$ DEEP

R1

VIEW II

$3\frac{1}{2}$

$1\frac{3}{4}$

$\frac{3}{4}$

$1\frac{1}{2}$

VIEW III

B

A C D E F G H I J K

L

VIEW I

U V W X

Z

$1\frac{3}{4}$

Y

VIEW II

T

M

N

O

S R Q P

VIEW III

MATERIAL: GRAY CAST IRON		
ORDER NO.	QUANTITY	
2A 56 - 72	24	
FEEDED PLATE		**BP-5**

ASSIGNMENT—UNIT 5: FEEDER PLATE (BP-5)

Student's Name _____

1. What is the name of the part?   1. _____

2. What is the blueprint number?   2. _____

3. What is the plate order number?   3. _____

4. How many parts are to be made?   4. _____

5. Name the material specified for the part.   5. _____

6. Study the Feeder Plate, BP-5.

    a.   Locate and name each line from Ⓐ to Ⓩ in the space provided for each in the table.

    b.   Tell how each line from Ⓐ to Ⓛ is identified. (NOTE: Line Ⓐ is filled in as a guide.)

Line on Drawing	(A) Name of Line	(B) How the Line is identified	
Ⓐ	*EXTENSION LINE*	*THIN, DARK, UNBROKEN LINE*	
Ⓑ			
Ⓒ			
Ⓓ			
Ⓔ			
Ⓕ			
Ⓖ			
Ⓗ			
Ⓘ			
Ⓙ			
Ⓚ			
Ⓛ			
Ⓜ		**Line**	**(A) Name of Line**
Ⓝ		Ⓤ	
Ⓞ		Ⓥ	
Ⓟ		Ⓦ	
Ⓠ		Ⓧ	
Ⓡ		Ⓨ	
Ⓢ		Ⓩ	
Ⓣ			

6   Three-View Drawings

Regularly shaped flat objects that require only simple processing operations are often adequately described with notes on a one-view drawing (see Unit 9). However, when the shape of the object changes, portions are cut away or relieved or complex machining or fabrication processes must be represented on a drawing, the one view may not be sufficient to describe the part accurately.

The number and selection of views is governed by the shape or complexity of the object. A view should not be drawn unless it makes a drawing easier to read or furnishes other information needed to describe the part clearly.

Throughout this text, as the student is required to interpret more complex drawings, the basic principles underlying the use of all additional views that are needed to describe the true shape of the object will be covered. Immediate application of these principles will then be made on typical industrial blueprints.

The combination of front, top, and right-side views represents the method most commonly used by draftspersons to describe simple objects. The manner in which each view is obtained and the interpretation of each view is discussed in this section.

THE FRONT VIEW

Before an object is drawn, it is examined to determine which views will best furnish the information required to manufacture the object. The surface that is to be shown as the observer looks at the object is called the front view. To draw this view, the draftsperson goes through an imaginary process of raising the object to eye level and turning it so that only one side can be seen. If an imaginary transparent plane is placed between the eye and the face of the object, parallel to the object, the image projected on the plane is the same as that formed in the eye of the observer, Figure 6–1.

Note in Figure 6–1 that the rays converge as they approach the observer's eye. If, instead of converging, these rays are parallel as they leave the object, the image they form on the screen is equivalent to a front view, as shown in Figure 6–2.

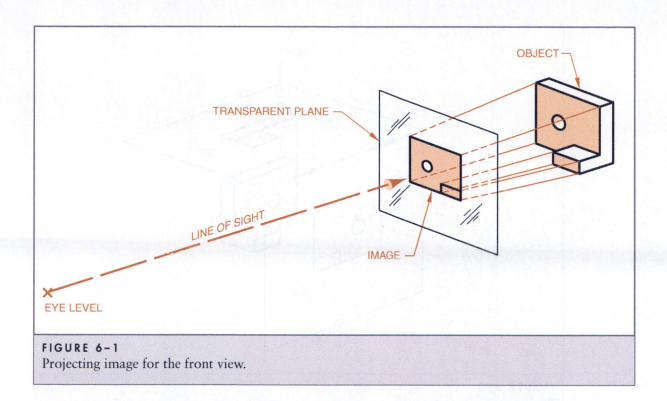

FIGURE 6-1
Projecting image for the front view.

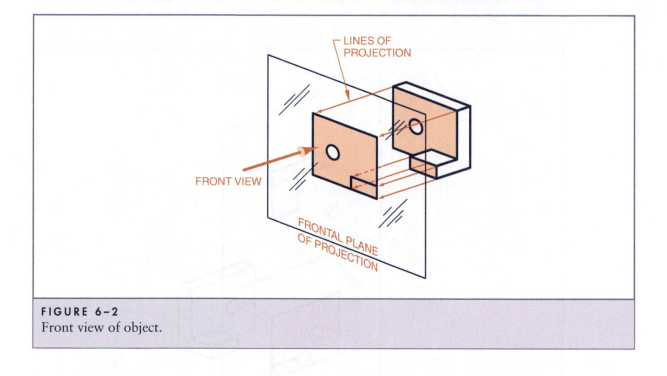

FIGURE 6-2
Front view of object.

THE TOP VIEW

To draw a **top view,** the draftsperson goes through a process similar to that required to obtain the **front view.** However, instead of looking squarely at the front of the object, the view is seen from a point directly above it, Figure 6–3.

When the horizontal plane on which the top view is projected is rotated so that it is in a vertical plane, as shown in Figure 6–4, the front and top views are in their proper relationship. In other words, the top view is always placed immediately above and in line with the front view.

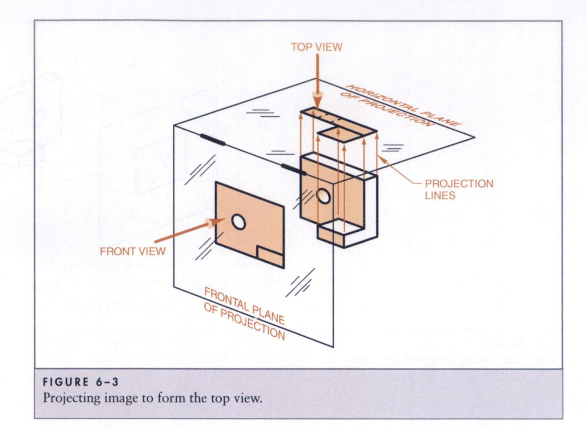

FIGURE 6-3
Projecting image to form the top view.

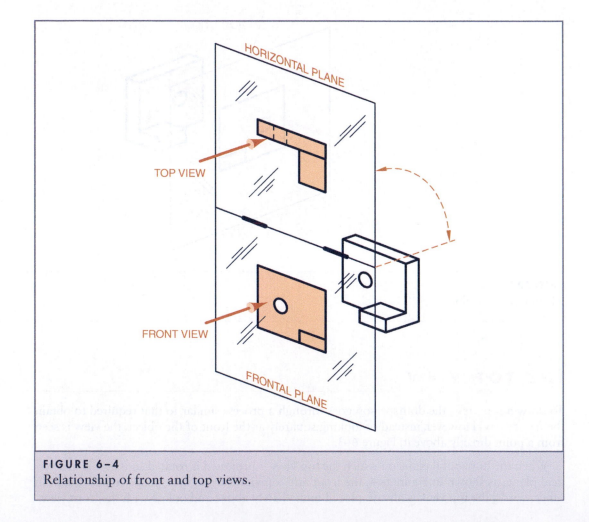

FIGURE 6-4
Relationship of front and top views.

THE SIDE VIEW

A side view is developed in much the same way that the other two views were obtained. That is, the draftsperson imagines the view of the object from the side that is to be drawn. This person then proceeds to draw the object as it would appear if parallel rays were projected upon a vertical plane, Figure 6–5.

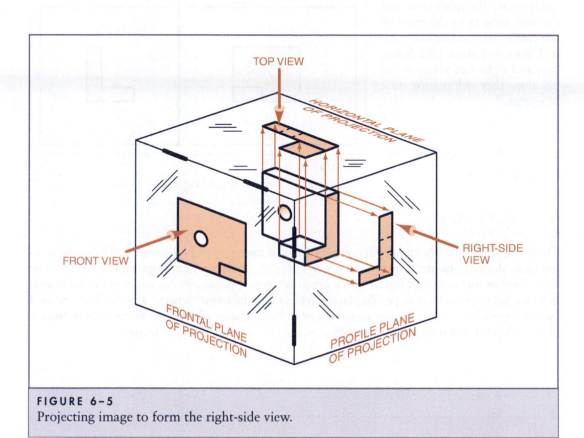

FIGURE 6–5
Projecting image to form the right-side view.

FRONT, TOP, AND RIGHT-SIDE VIEWS

By swinging the top of the imaginary projection box to a vertical position and the right-side forward, the top view is directly above the front view and the side view is to the right of the front view and in line with it. Figure 6–6 shows the front, top, and right-side views in the positions they will occupy on a blueprint.

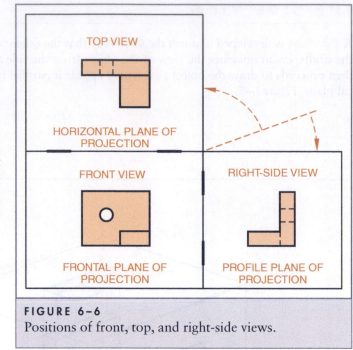

FIGURE 6–6
Positions of front, top, and right-side views.

HEIGHT, WIDTH (LENGTH), AND DEPTH DIMENSIONS

The terms height, width, and depth refer to specific dimensions or part sizes. ASME designations for these dimensions are shown in Figure 6–7. **Height** is the vertical distance between two or more lines or surfaces (part features) that are in horizontal planes. **Width** refers to the horizontal distance between surfaces in profile planes. In the shop, the terms "length" and "width" are used interchangeably. **Depth** is the horizontal (front to back) distance between two features in frontal planes. Depth is often identified in the shop as the **thickness** of a part or feature.

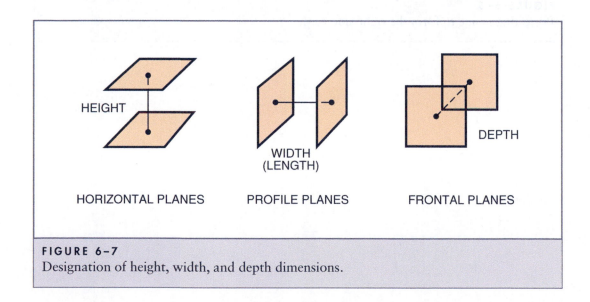

FIGURE 6–7
Designation of height, width, and depth dimensions.

WORKING DRAWINGS

An actual drawing of a part shows only the top, front, and right-side views without the imaginary transparent planes, Figure 6–8. These views show the exact shape and size of the object and define the relationship of one view to another.

A drawing, when completely dimensioned and with necessary notes added, is called a **working drawing** because it furnishes all the information required to construct the object, Figure 6–9.

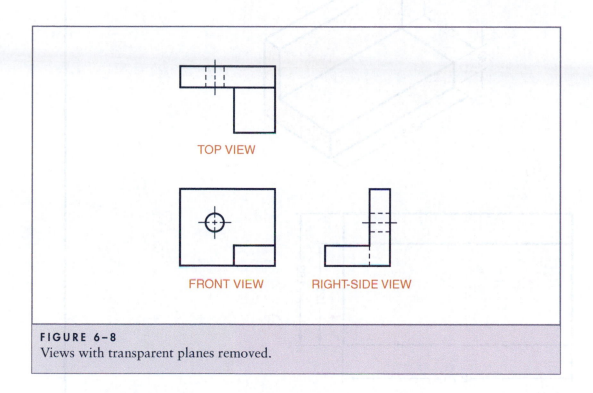

TOP VIEW

FRONT VIEW   RIGHT-SIDE VIEW

FIGURE 6–8
Views with transparent planes removed.

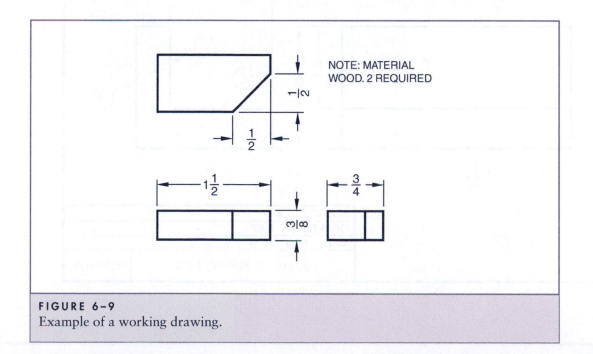

NOTE: MATERIAL
WOOD. 2 REQUIRED

$\frac{1}{2}$

$\frac{1}{2}$

$1\frac{1}{2}$

$\frac{3}{8}$

$\frac{3}{4}$

FIGURE 6–9
Example of a working drawing.

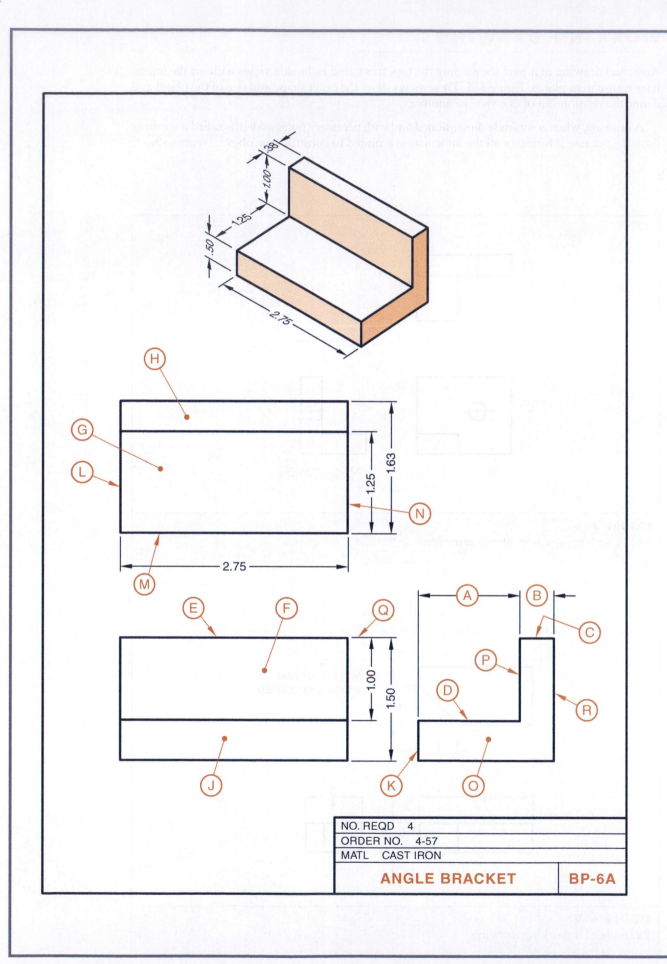

NO. REQD	4
ORDER NO.	4-57
MATL	CAST IRON

ANGLE BRACKET | **BP-6A**

ASSIGNMENT A—UNIT 6: ANGLE BRACKET (BP-6A)

Student's Name _____

1. How many angle brackets are required?   1._____

2. Name the material specified for the angle bracket.   2._____

3. State the order number of the angle bracket.   3._____

4. What is the overall width (length) of the angle bracket?   4._____

5. What is the overall height?   5._____

6. What is the overall depth?   6._____

7. What is dimension Ⓐ?   7._____

8. What is dimension Ⓑ?   8._____

9. What surface in the top view is represented by line Ⓒ in the right-side view?   9._____

10. Name the three views that are used to describe the shape and size of the part.   10. (a) _____

   (b) _____

   (c) _____

11. What surface in the top view is represented by line Ⓓ in the right-side view?   11._____

12. What line in the right-side view represents surface Ⓕ in the front view?   12._____

13. What line in the right-side view represents surface Ⓙ in the front view?   13._____

14. What line in the top view represents surface Ⓞ in the right-side view?   14._____

15. What line in the front view represents surface Ⓗ in the top view?   15._____

16. What line in the right-side view represents surface Ⓗ in the top view?   16._____

17. What kind of lines are Ⓒ, Ⓓ, Ⓔ, Ⓚ, and Ⓛ?   17._____

18. What kinds of lines are Ⓐ and Ⓑ?   18._____

19. What encircled letter denotes an extension line?   19._____

20. What encircled letter in the front view denotes an object line?   20._____

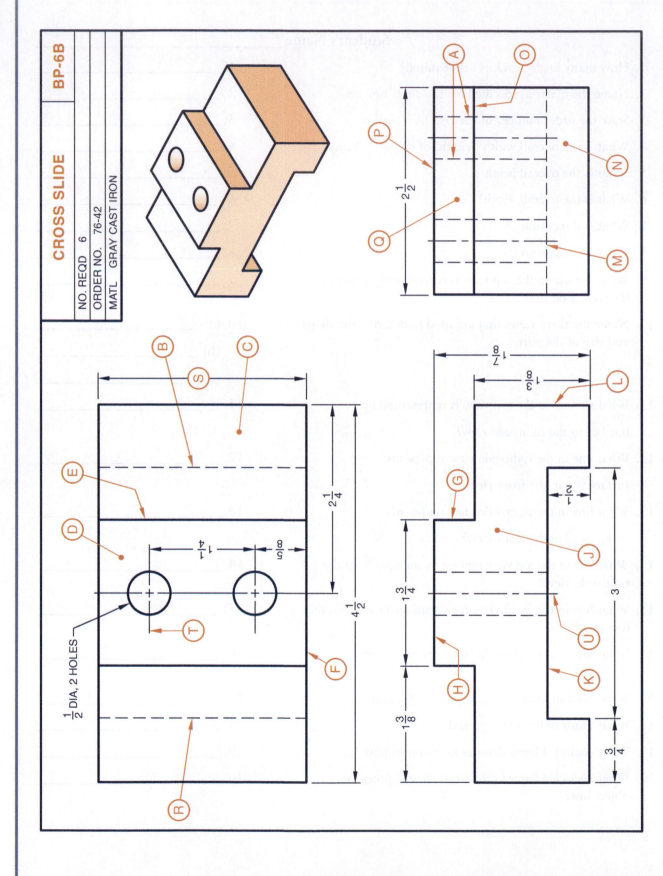

CROSS SLIDE

BP-6B

NO. REQD 6

ORDER NO. 76-42

MATL GRAY CAST IRON

$2\frac{1}{2}$

$1\frac{7}{8}$

$1\frac{3}{8}$

$\frac{1}{2}$

$1\frac{3}{4}$

3

$1\frac{3}{8}$

$\frac{3}{4}$

$2\frac{1}{4}$

$4\frac{1}{2}$

$1\frac{1}{4}$

$\frac{5}{8}$

$\frac{1}{2}$ DIA, 2 HOLES

ASSIGNMENT B—UNIT 6: CROSS SLIDE (BP-6B)

Student's Name _____

1. What material is used for the cross slide?

2. How many pieces are required?

3. What is the overall width (length) of the cross slide?

4. What is the order number?

5. What is the overall height of the cross slide?

6. What are the lines marked Ⓐ and Ⓑ called?

7. What do the lines marked Ⓐ represent?

8. What two lines in the top view represent the slot shown in the front view?

9. What line in the right-side view represents the slot shown in the front view?

10. What line in the front view represents surface Ⓠ in the right-side view?

11. What line in the front view represents surface Ⓓ in the top view?

12. What line in the top view represents surface Ⓙ in the front view?

13. What line in the side view represents surface Ⓓ in the top view?

14. What is the diameter of the holes?

15. What is the center-to-center dimension of the holes?

16. How far is the center of the first hole from the front surface of the side?

17. Are the holes drilled all the way through the slide?

18. What is the width of the slot shown in the front view?

19. What is the height of the slot?

20. Determine dimension Ⓢ.

21. What is the width of the projection at the top of the slide?

22. How high is the projection?

23. What kind of line is used at Ⓞ and Ⓟ?

24. What kind of line is used at Ⓜ?

1. _____
2. _____
3. _____
4. _____
5. _____
6. _____
7. _____
8. _____
9. _____
10. _____
11. _____
12. _____
13. _____
14. _____
15. _____
16. _____
17. _____
18. _____
19. _____
20. _____
21. _____
22. _____
23. _____
24. _____

7

Arrangement of Views

The main purpose of a drawing is to give the technician sufficient information needed to build, inspect, or assemble a part or mechanism according to the specifications of the designer. Since the selection and arrangement of views depends on the complexity of a part, only those views should be drawn that help in the interpretation of the drawing.

The average drawing that includes front, top, and side views is known as a three-view drawing. However, the designation of the views is not as important as the fact that the combination of views must give all the details of construction in the most understandable way.

The draftsperson usually selects as a front view of the object that view which best describes the general shape of the part. This front view may have no relationship to the actual front position of the part as it fits into a mechanism.

The names and positions of the different views that may be used to describe an object are illustrated in Figure 7–1. Note that the back view may be placed in any one of three locations. The views that are easiest to read and, at the same time, furnish all the required information should be the views selected for the drawing.

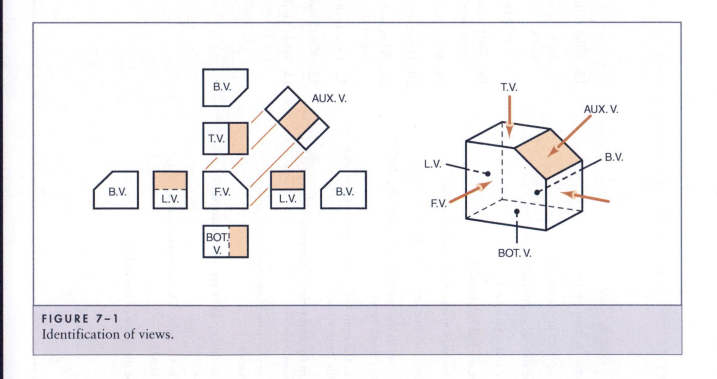

FIGURE 7–1
Identification of views.

The name and abbreviation for each view is identified throughout this text as shown in Figure 7–2.

Name of View	Abbreviation
Front View	(F.V.)
Right-Side View	(R.V.)
Left-Side View	(L.V.)
Bottom View	(Bot. V.)
Back or Rear View	(B.V.)
Auxiliary View	(Aux. V.)
Top View	(T.V.)

FIGURE 7–2
Abbreviations of different views.

ASSIGNMENT—UNIT 7: BLOCK SLIDE (BP-7)

Student's Name _____

Study the pictorial drawing of the slide base (Figure 7–3) with reference to the front view. Then, identify and place the name of each view in the spaces provided in Figure 7–4 and Figure 7–5.

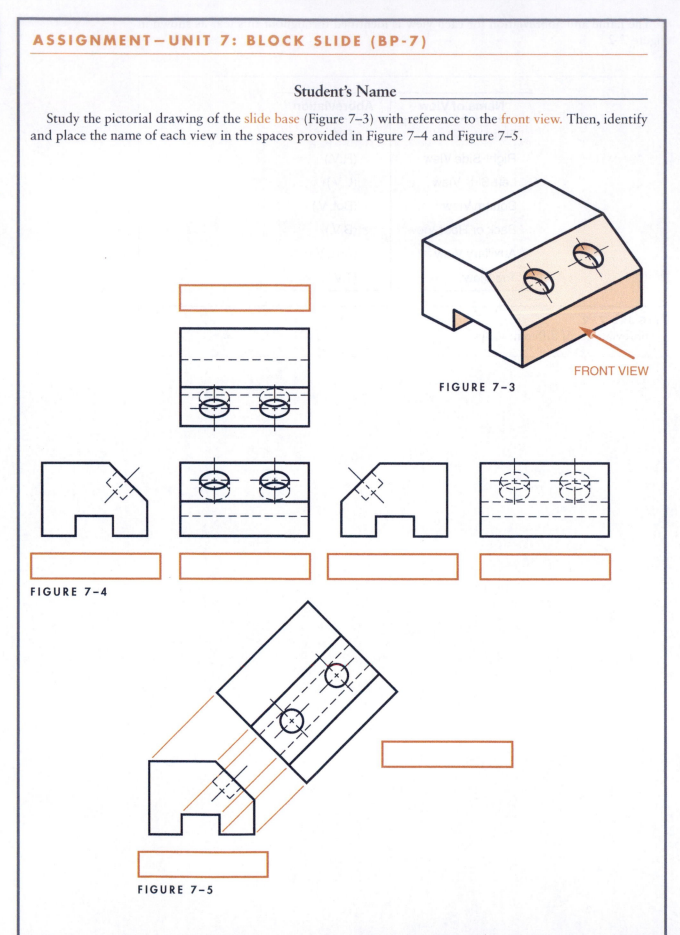

FIGURE 7–3

FRONT VIEW

FIGURE 7–4

FIGURE 7–5

Two-View Drawings

Simple, symmetrical (SYMM) flat objects and symmetrical cylindrical parts, such as sleeves, shafts, rods, or studs, require only two views to show the full details of construction, Figure 8–1. The two views usually include the front view and a right-side or left-side view, or a top or bottom view.

Symmetrical (SYMM) means that features on both sides of a center line Ꞔ shown on a drawing are the same size and shape. Objects are indicated by two equal-length short, dark, parallel lines. These lines are placed outside the drawing of the object on its center line Ꞔ, Figure 8–1 below and BP-8C, page 46.

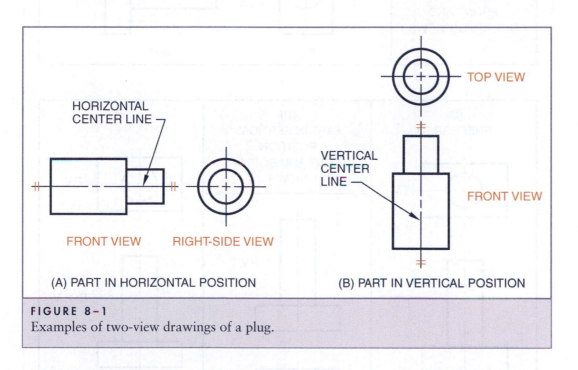

FIGURE 8–1
Examples of two-view drawings of a plug.

In the front view, Figure 8–1, the center line runs through the axis of the part as a horizontal center line. If the plug is in a vertical position, the center line runs through the axis as a vertical center line.

The second view of the two-view drawing contains a horizontal and a vertical center line intersecting at the center of the circles that make up the part in this view. Combinations of views commonly used in industrial blueprints are shown in Figure 8–2.

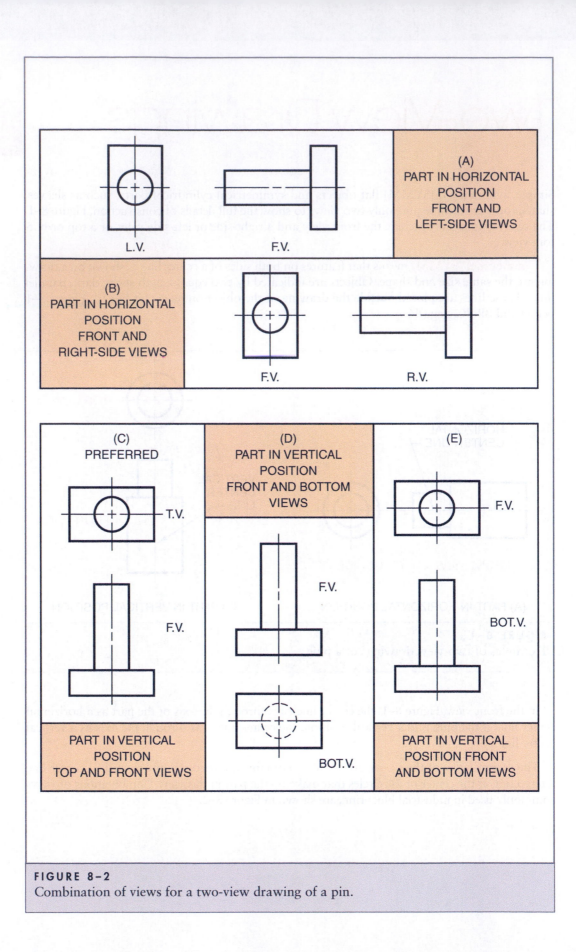

FIGURE 8–2
Combination of views for a two-view drawing of a pin.

Note in Figure 8–2 that different names are used to identify the same views. There are two main reasons why a draftsperson will show a **primary view,** also known as the front view. Either this view shows

a. the object's normal operating position or

b. it shows something unique about its shape.

A hidden detail may be straight, slanted curved, or cylindrical. Whatever the shape of the detail and regardless of the number or positions of views, the hidden detail is represented by a hidden edge or invisible edge line, Figure 8–3.

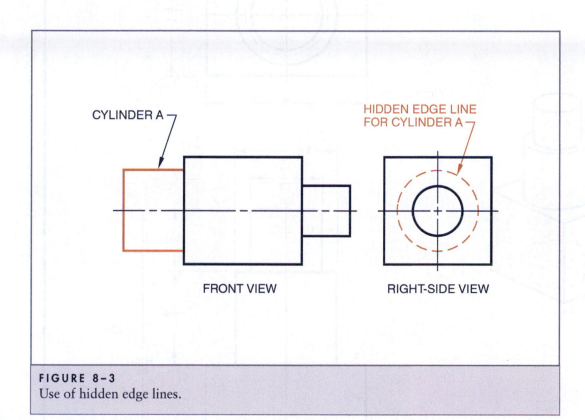

CYLINDER A

HIDDEN EDGE LINE
FOR CYLINDER A

FRONT VIEW

RIGHT-SIDE VIEW

FIGURE 8–3
Use of hidden edge lines.

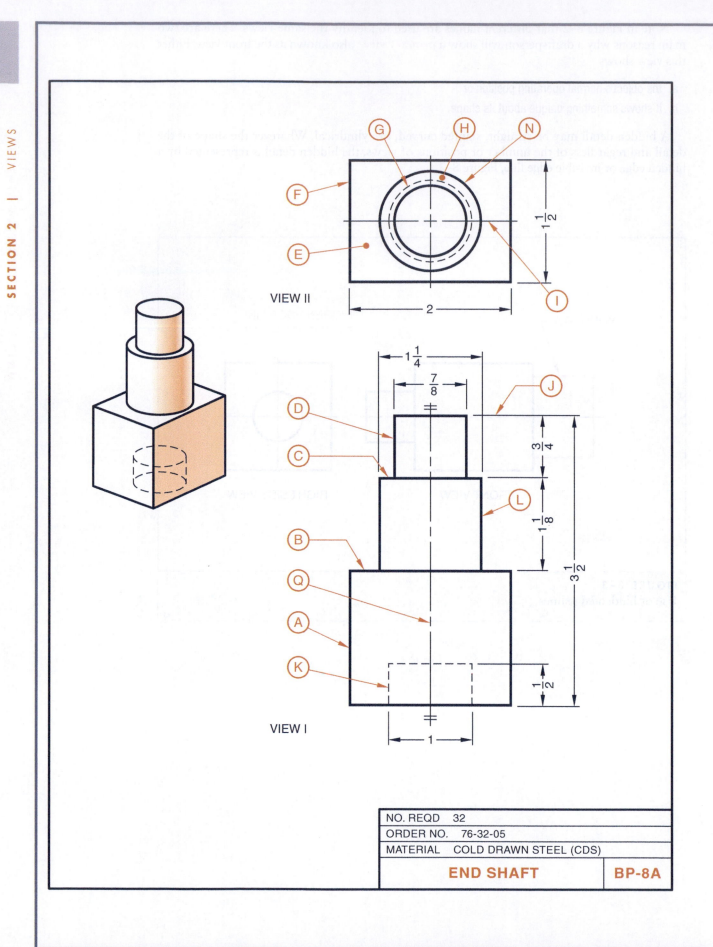

G H N

F

E

VIEW II

$1\frac{1}{2}$

2

I

$1\frac{1}{4}$

$\frac{7}{8}$

J

D

C

L

$\frac{3}{4}$

$1\frac{1}{8}$

$3\frac{1}{2}$

B

Q

A

K

$1\frac{1}{2}$

VIEW I

1

NO. REQD	32	
ORDER NO.	76-32-05	
MATERIAL	COLD DRAWN STEEL (CDS)	
END SHAFT		**BP-8A**

ASSIGNMENT A—UNIT 8: END SHAFT (BP-8A)

Student's Name _____

1. Name the two views shown.   1. _____ _____

2. What line in VIEW II represents surface (A)?   2. _____

3. What lettered surface in VIEW II represents surface (B)?   3. _____

4. What circle in VIEW II represents the 1″ hole?   4. _____

5. What line in VIEW 1 represents surface (H)?   5. _____

6. Name line (I).   6. _____

7. What kind of line is (D)?   7. _____

8. Name line (J).   8. _____

9. What kind of line is (K)?   9. _____

10. What circle in the top view represents diameter (L)?   10. _____

11. What letters in VIEW I represent object lines?   11. _____

12. What letters in VIEWS I and II represent center lines?   12. _____

13. Give the diameter of (L).   13. _____

14. What is the smallest diameter of the shaft?   14. _____

15. Determine the height (length) of the 1¼″ diameter portion.   15. _____

16. What is the height (length) of the rectangular part of the shaft?   16. _____

17. Is the end shaft SYMM?   17. _____

18. Give the overall height (length) of the shaft.   18. _____

19. What is the order number?   19. _____

20. State the material from which the shaft is to be machined.   20. _____

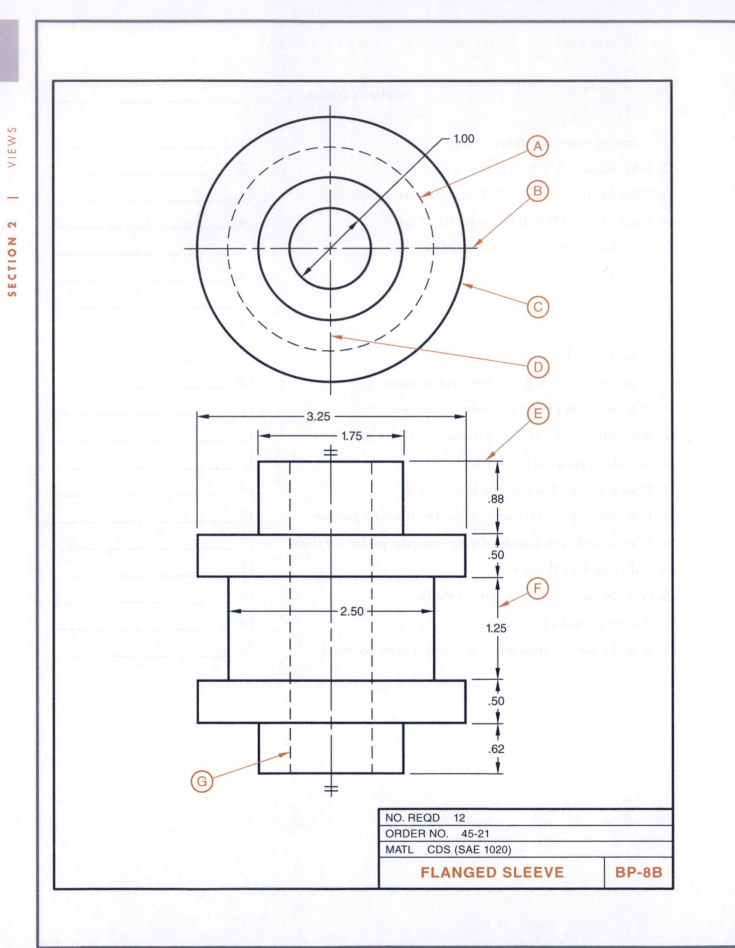

1.00

Ⓐ

Ⓑ

Ⓒ

Ⓓ

Ⓔ

3.25

1.75

.88

.50

Ⓕ

2.50

1.25

.50

.62

Ⓖ

NO. REQD	12	
ORDER NO.	45-21	
MATL	CDS (SAE 1020)	
FLANGED SLEEVE		**BP-8B**

ASSIGNMENT B—UNIT 8: FLANGED SLEEVE (BP-8B)

Student's Name _____

1. Is this part SYMM?   1. _____

2. What is the order number?   2. _____

3. How many pieces are required?   3. _____

4. What material is used?   4. _____

5. Name the two views that are used to represent the flanged sleeve.   5. _____

6. Name the kind of line indicated by each of the following encircled letters.   _____

    Ⓐ   6. Ⓐ _____

    Ⓑ   Ⓑ _____

    Ⓒ   Ⓒ _____

    Ⓓ   Ⓓ _____

    Ⓔ   Ⓔ _____

    Ⓕ   Ⓕ _____

    Ⓖ   Ⓖ _____

7. What is the outside diameter of both flanges?   7. _____

8. What is the height (thickness) of each flange?   8. _____

9. What is the diameter of the center hole?   9. _____

10. Does the hole go all the way through the center of the sleeve?   10. _____

11. What is the diameter of the hidden circle?   11. _____

12. Determine the total or overall height of the flanged sleeve.   12. _____

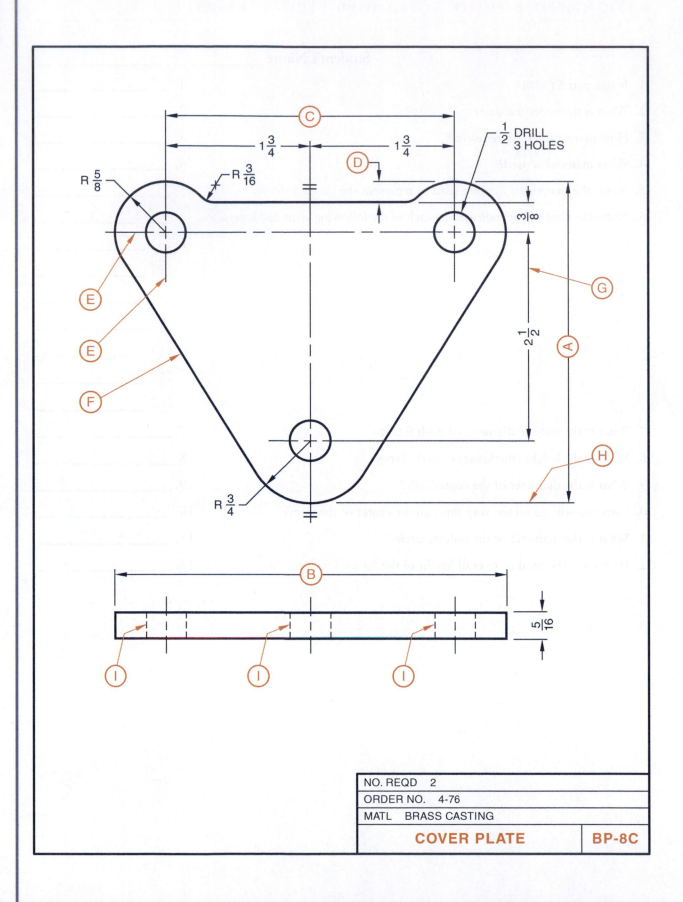

C

$1\frac{3}{4}$ $1\frac{3}{4}$

D

$\frac{1}{2}$ DRILL
3 HOLES

$R\frac{5}{8}$ $R\frac{3}{16}$

$\frac{3}{8}$

E

G

E

$2\frac{1}{2}$ A

F

H

$R\frac{3}{4}$

B

$\frac{5}{16}$

I   I   I

NO. REQD	2
ORDER NO.	4-76
MATL	BRASS CASTING

COVER PLATE **BP-8C**

ASSIGNMENT C—UNIT 8: COVER PLATE (BP-8C)

Student's Name _____

1. Name the material specified for the part.

   1. _____

2. Name the two views used to describe the part.

   2. _____

      _____

3. Identify the kind of line indicated by each of the following encircled letters.

    Ⓔ

    Ⓕ

    Ⓖ

    Ⓗ

    Ⓘ

   3. Ⓔ _____

      Ⓕ _____

      Ⓖ _____

      Ⓗ _____

      Ⓘ _____

4. What is the overall depth Ⓐ?

   4. _____

5. What is the overall length Ⓑ?

   5. _____

6. How many holes are to be drilled?

   6. _____

7. What is the thickness (height) of the plate?

   7. _____

8. What is the diameter of the holes?

   8. _____

9. What is the distance between the center of one of the two upper holes and the center line of the cover plate?

   9. _____

10. Give the center distance Ⓒ of the two upper holes.

   10. _____

11. What is the radius that forms the two upper rounds of the cover plate?

   11. _____

12. What radius forms the lower part of the cover plate?

   12. _____

13. What kind of line is drawn through the center of the cover plate?

   13. _____

14. Determine distance Ⓓ.

   14. _____

15. How much stock is left between the edge of one of the upper holes and the outside of the piece?

   15. _____

One-View Drawings

One-view drawings are commonly used in the industry to represent parts that are uniform in shape. These drawings are often supplemented with **notes,** symbols, and written information, Figure 9–1B and Figure 9–3. They are normally used to describe the shape of cylindrical, cone-shaped, rectangular, and other symmetrical parts. **Leaders** are used to relate a note to a particular feature, as in Figure 9–2.

Thin, flat objects of uniform thickness are represented by one-view drawings. The one-view drawing at (A) in Figure 9–1 represents a cylinder of a given length and diameter. The drawing at (B) represents a symmetrical flat part.

SYMBOLS, ABBREVIATIONS, AND NOTES ON ONE-VIEW DRAWINGS

In the two examples in Figure 9–1, the use of the abbreviation **DIA** in (A), and the **Ø** and **R** symbols with the thickness note in (B) take the place of second views. A capital R always refers to a radius on a blueprint. The letters "DIA" refer to the diameter of cylindrical objects. However, diameters are most commonly shown by using the symbol **Ø,** which will precede the actual numeric dimension given on a drawing. Example: In Figure 9–3, the diameter of the two holes in the gasket is (.375), shown as **Ø .375.** Refer to page 49, Figure 9–3 and Figure 9–4 and drawing BP-9A on page 51. These will also show the usage of abbreviations and symbols.

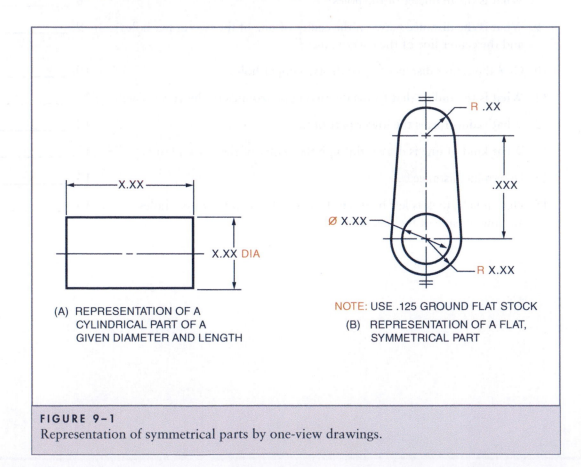

(A) REPRESENTATION OF A CYLINDRICAL PART OF A GIVEN DIAMETER AND LENGTH

(B) REPRESENTATION OF A FLAT, SYMMETRICAL PART

NOTE: USE .125 GROUND FLAT STOCK

FIGURE 9–1
Representation of symmetrical parts by one-view drawings.

Figure 9–2 shows the use of the abbreviations HEX and SQ to indicate the hexagon-shaped head and the square form for the opposite end. The combination of the one view and the supplemental information indicates the millimeter sizes for the hexagon head, the body diameters and lengths, the dimensions of the square end, and the overall length. A pictorial drawing, which normally would not be included, shows what the part looks like.

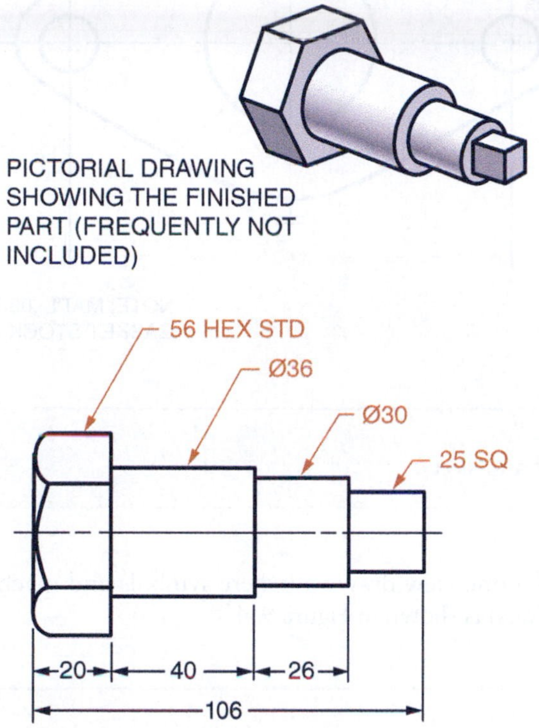

PICTORIAL DRAWING
SHOWING THE FINISHED
PART (FREQUENTLY NOT
INCLUDED)

56 HEX STD
Ø36
Ø30
25 SQ

20   40   26
106

FIGURE 9–2
Use of leaders, abbreviations, and symbols on a one-view drawing.

This technique of providing what would be third dimensions (on a second or third view) in the form of supplemental notes on a one-view drawing is widely used for drawings of gaskets, shims, and other thin, flat plates. Figure 9–3 is a typical one-view drawing of a gasket. Since the third dimension, such as thickness (depth), can be given as a note, gaskets, shims, and the like may be represented on a drawing by one view.

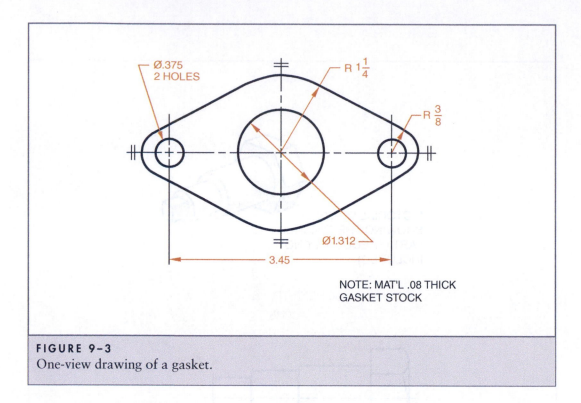

FIGURE 9–3
One-view drawing of a gasket.

Another example of a one-view drawing (where symbols and machining notes are used to supplement the dimensions) is shown in Figure 9–4.

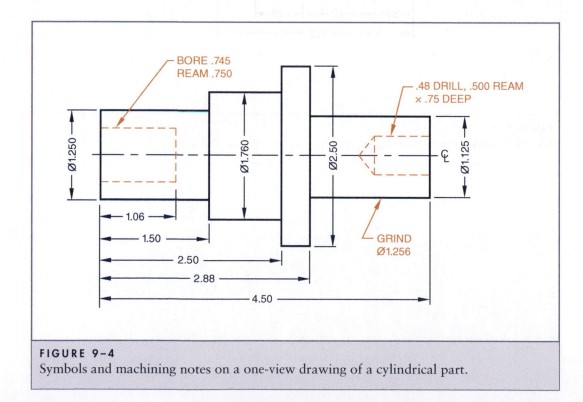

FIGURE 9–4
Symbols and machining notes on a one-view drawing of a cylindrical part.

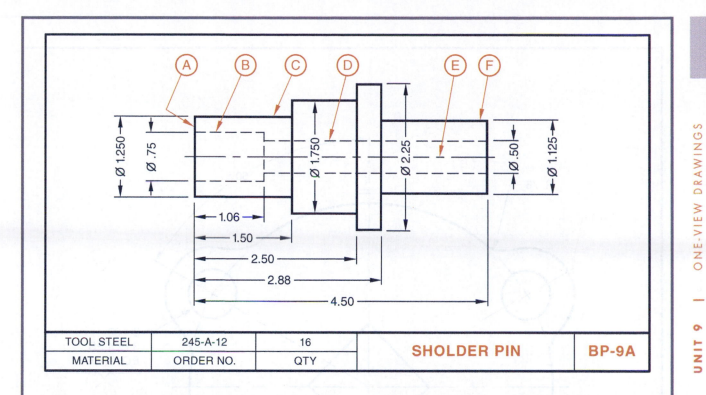

TOOL STEEL	245-A-12	16	SHOLDER PIN	BP-9A
MATERIAL	ORDER NO.	QTY		

ASSIGNMENT A—UNIT 9: SHOULDER PIN (BP-9A)

Student's Name _____

1. Name the view represented on BP-9A. 1. _____

2. What is the shape of the shoulder pin? 2. _____

3. How many outside diameters are shown? 3. _____

4. What is the largest diameter? 4. _____

5. What diameter is the smallest hole? 5. _____

6. What is the overall length of the pin? 6. _____

7. How deep is the .75″ hole? 7. _____

8. How wide is the Ø 1.750″ portion? 8. _____

9. What letters represent object lines? 9. _____

10. What kind of lines are B and D? 10. _____

11. What letter represents the center line? 11. _____

12. What does the center line indicate about the holes and outside diameters? 12. _____

13. Give the thickness of the Ø 2.25″ portion. 13. _____

14. State the order number of the part. 14. _____

15. What material is specified for the pins? 15. _____

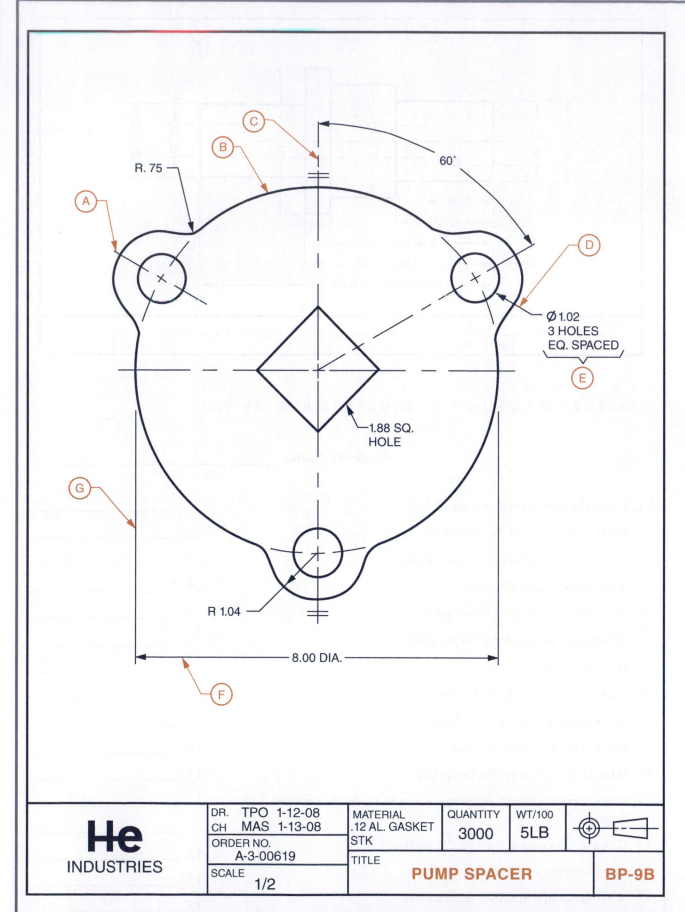

R. 75

A

B

C

60°

D

Ø 1.02
3 HOLES
EQ. SPACED

E

1.88 SQ.
HOLE

G

R 1.04

8.00 DIA.

F

He INDUSTRIES	DR. TPO 1-12-08 CH MAS 1-13-08	MATERIAL .12 AL. GASKET STK	QUANTITY 3000	WT/100 5LB
	ORDER NO. A-3-00619	TITLE **PUMP SPACER**		
	SCALE 1/2			**BP-9B**

ASSIGNMENT B—UNIT 9: PUMP SPACER (BP-9B)

Student's Name _____

1. Name the type of drawing used to represent the pump spacer.

2. Explain briefly why all information required to produce the part may be provided in one view.

3. Tell why the part is symmetrical with the vertical axis.

4. Identify the:

   (a) Order Number

   (b) Print Number

5. Indicate the following:

   (a) The thickness and kind of material in the pump spacer.

   (b) The quantity to be produced.

6. Calculate the weight of the stamped parts in the order.

7. Identify the letter that relates to each of the following types of lines or function:

   (a) Object Lines

   (b) Center Lines

   (c) Extension Lines

   (d) Dimension Lines

   (e) Leaders

8. Specify the angular dimension between each of the three center lines for the three equally spaced holes on the pump spacer.

9. Give the size of the square hole.

10. Calculate the number of degrees center line Ⓐ is from center line Ⓒ.

11. Explain the meaning of the note Ⓔ.

12. Compute the radius to the center of the 1.02 holes.

13. Determine the following dimensions:

    (a) The body diameter of the pump spacer.

    (b) The radius of the lugs.

    (c) The overall outside diameter.

1. _____

2. _____
   _____

3. _____
   _____

4. (a) Order Number _____

   (b) Print Number _____

5. Weight _____

   (a) _____

   (b) Number of Parts _____

6. _____

7. _____

   (a) Object Lines _____

   (b) Center Lines _____

   (c) Extension Lines _____

   (d) Dimension Lines _____

   (e) Leaders _____

8. _____

9. _____

10. _____

11. _____
    _____
    _____
    _____

12. Ⓡ = _____

13. (a) Body Diameter _____

    (b) Lug Radius _____

    (c) Overall Outside Diameter _____
    _____

10 Auxiliary Views

As long as all the surfaces of an object are parallel or at right angles to one another, they may be represented in one or more regular views, Figure 10–1.

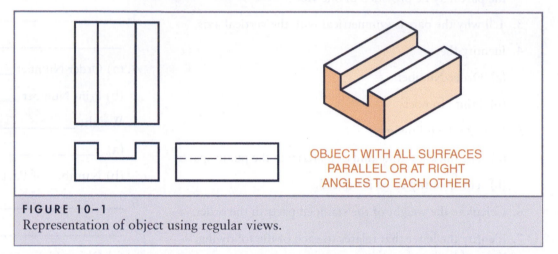

OBJECT WITH ALL SURFACES
PARALLEL OR AT RIGHT
ANGLES TO EACH OTHER

FIGURE 10–1
Representation of object using regular views.

The surfaces of such objects can be projected in their true sizes and shapes on either a horizontal or a vertical plane or on any combination of these planes.

Some objects have one or more surfaces that slant and are inclined away from either a horizontal or vertical plane. In this situation, the regular views will not show the true shape of the inclined surface, Figure 10–2. If the true shape must be shown, the drawing must include **an auxiliary view** to represent the angular surface accurately. The auxiliary view is in addition to the regular views.

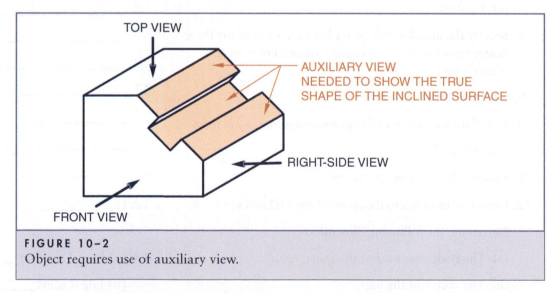

TOP VIEW

AUXILIARY VIEW
NEEDED TO SHOW THE TRUE
SHAPE OF THE INCLINED SURFACE

RIGHT-SIDE VIEW

FRONT VIEW

FIGURE 10–2
Object requires use of auxiliary view.

APPLICATION OF AUXILIARY VIEWS

Auxiliary views may be full views or partial views. Figure 10–3 shows an auxiliary view in which only the inclined surface and other required details are included. In an auxiliary view, the inclined surface is projected on an imaginary plane that is parallel to it. Rounded surfaces and circular holes, which are distorted and appear as ellipses in the regular views, will appear in their true shapes and sizes in an auxiliary view.

Auxiliary views are named according to the position from which the inclined face is seen. For example, the auxiliary view may be an auxiliary front, top, bottom, left-side, or right-side view. On drawings of complex parts involving compound angles, one auxiliary view may be developed from another auxiliary view. The first auxiliary view is called the primary view, and those views developed from it are called secondary auxiliary views. For the present, attention is focused on primary auxiliary views.

SUMMARY

Auxiliary views are usually partial views that show only the inclined surface of an object. In Figure 10–3, the true size and shape of surface Ⓐ is shown in the auxiliary view of the angular surface.

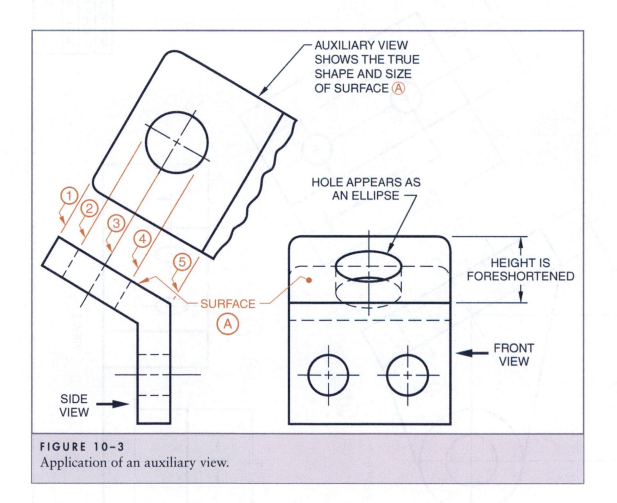

FIGURE 10–3
Application of an auxiliary view.

The draftsperson develops this view by projecting lines ①, ②, ③, ④, and ⑤ at right angles to surface Ⓐ. When the front view is compared with the partial auxiliary view, it can be seen that the hole in surface Ⓐ appears as an ellipse in the front view. The height of surface Ⓐ is foreshortened in the regular view, while the true shape and size are shown on the auxiliary view.

In Figure 10–3, the combination of left-side view, partial auxiliary top view, and front view, when properly dimensioned, shows all surfaces in their true size and shape. As a result, these are the only views required to describe this part completely.

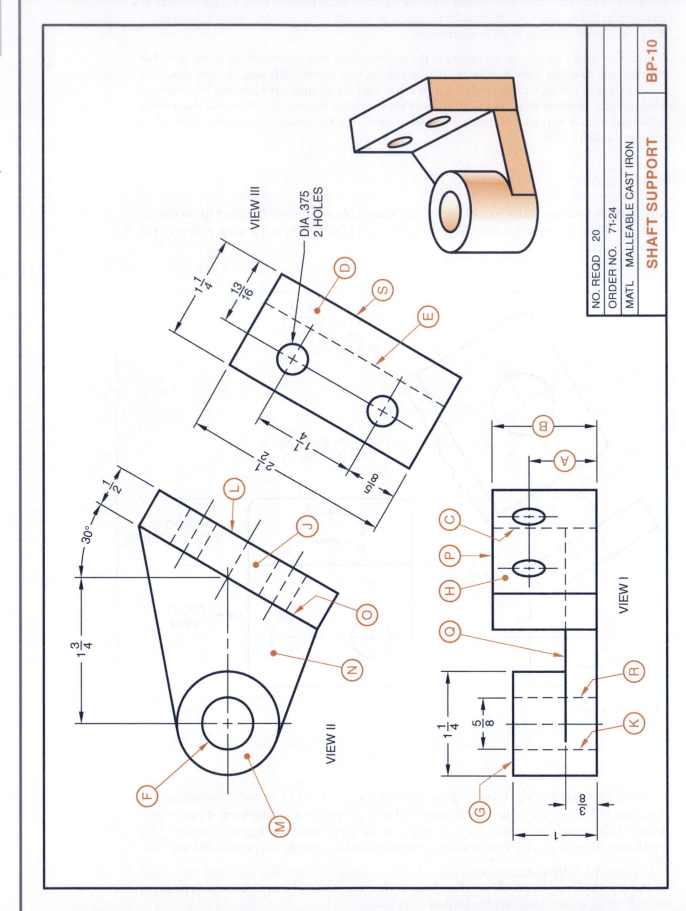

VIEW III

DIA .375
2 HOLES

1 1/4

13/16

D

S

E

2 1/2

1 1/4

3/8

VIEW II

1/2

30°

1 3/4

L

J

O

N

F

M

B

A

C

P

H

Q

R

K

G

1 1/4

5/8

3/8

1

VIEW I

NO. REQD	20
ORDER NO.	71-24
MATL	MALLEABLE CAST IRON

SHAFT SUPPORT

BP-10

ASSIGNMENT—UNIT 10 SHAFT SUPPORT (BP-10)

Student's Name _____

1. Name the material specified for the part.

2. How many pieces are required?

3. What is the order number?

4. Name each of the following:

   VIEW I

   VIEW II

   VIEW III

5. What kind of line is (C)?

6. What kind of line is (P)?

7. Give the diameter of hole (F).

8. Determine dimension (A).

9. Determine dimension (B).

10. What surface in the front view is the same as surface (D) in the auxiliary view?

11. How many DIA .375" holes must be made in the support?

12. What is the center-to-center distance of the DIA .375" holes?

13. What surface in the top view is represented by the line (E) in the auxiliary view?

14. What surface in the top view is represented by line (G) in the front view?

15. What line in the top view represents surface (H) in the front view?

16. What line in the front view represents surface (J) in the top view?

17. What do (K) lines and (R) represent?

18. What line in the front view represents surface (N) in the top view?

19. Determine the vertical distance between the surface represented by the line (P) and the surface represented by line (Q).

20. What is the distance between the face represented by line (E) and the surface represented by line (S)?

1. _____

2. _____

3. _____

4. (I) _____

   (II) _____

   (III) _____

5. _____

6. _____

7. _____

8. _____

9. _____

10. _____

11. _____

12. _____

13. _____

14. _____

15. _____

16. _____

17. _____

18. _____

19. _____

20. _____

Dimensions and Notes

11

Size and Location Dimensions

Drawings consist of several types of lines that are used singly or in combination with each other to describe the shape and internal construction of an object or mechanism. However, to construct or machine a part, the blueprint or drawing must include dimensions that indicate exact sizes and locations of surfaces, indentations, holes, and other details.

The lines and dimensions, in turn, are supplemented by notes that give additional information. This information includes the kind of material used, the degree of machining accuracy required, details regarding the assembly of parts, and any other data that the craftsperson needs to know to make and assemble the part.

THE LANGUAGE OF DRAFTING

To ensure some measure of uniformity in industrial drawings, the American Society of Mechanical Engineers (ASME) has published drafting standards. These standards are called the language of drafting and are in general use throughout North America. While these drafting standards or practices may vary in some respects between industries, the principles are basically the same. The practices recommended by ASME for dimensioning and for making notes are generally followed in this section.

Standards for Dimensioning

All drawings should be dimensioned completely so that a minimum of computation is necessary and the parts can be built without scaling the drawing. However, there should not be a duplication of dimensions unless such dimensions make the drawing clearer and easier to read.

Many parts cannot be drawn full size because they are too large to fit a standard drawing sheet or computer screen or too small to have all details shown clearly. The draftsperson can, however, still represent such objects either by reducing (in the case of large objects) or enlarging (for small objects) the size to which the drawing is made.

This practice does not affect any dimensions, as the dimensions on a drawing give the actual sizes. If a drawing 6″ long represents a part 12″ long, a note should appear in the title box of the drawing to indicate the scale that is used. This scale is the ratio of the drawing size to the actual size of the object. In this case, the scale 6″ = 12″ is called half scale, or 1:2. Other common scales include the one-quarter scale (1:4), one-eighth scale (1:8), and the double-size scale (2:1).

CONSTRUCTION DIMENSIONS

Dimensions used in building a part are sometimes called **construction dimensions.** These dimensions serve two purposes: (1) they indicate size, (S), and (2) they give exact locations (L). For example, to drill a through hole in a part, the technician must know the diameter of the hole and the exact location of the center of the hole, Figure 11–1.

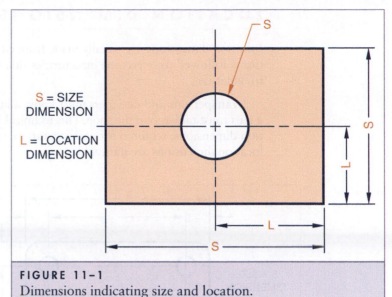

S = SIZE DIMENSION

L = LOCATION DIMENSION

FIGURE 11–1
Dimensions indicating size and location.

TWO-PLACE DECIMAL DIMENSIONS

Dimensions may appear on drawings as **two-place decimals.** These are widely used when the range of dimensional accuracy of a part is between 0.01″ larger or smaller than a specified dimension (basic size). Where possible, two-place decimal dimensions are given in even hundredths of an inch.

Three- and four-place decimal dimensions continue to be used for more precise dimensions requiring machining accuracies in thousandths or ten-thousandths of an inch.

SIZE DIMENSIONS

Every solid has three size dimensions: depth or thickness, length or width, and height. In the case of a regular prism, two of the dimensions are usually placed on the principal view and the third dimension is placed on one of the other views, Figure 11–2.

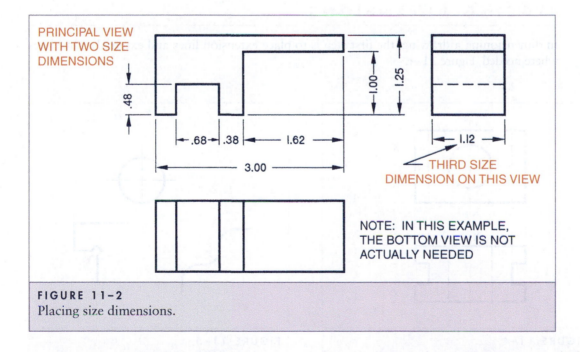

PRINCIPAL VIEW WITH TWO SIZE DIMENSIONS

THIRD SIZE DIMENSION ON THIS VIEW

NOTE: IN THIS EXAMPLE, THE BOTTOM VIEW IS NOT ACTUALLY NEEDED

FIGURE 11–2
Placing size dimensions.

LOCATION DIMENSIONS

Location dimensions are usually made from either a center line or a finished surface. This practice is followed to overcome inaccuracies due to variations in measurement caused by surface irregularities.

Draftspersons and designers must know all the manufacturing processes and operations that a part must undergo in the shop. This technical information assists them in placing size and location dimensions, required for each operation, on the drawing. Figure 11–3 shows how size and location dimensions are indicated.

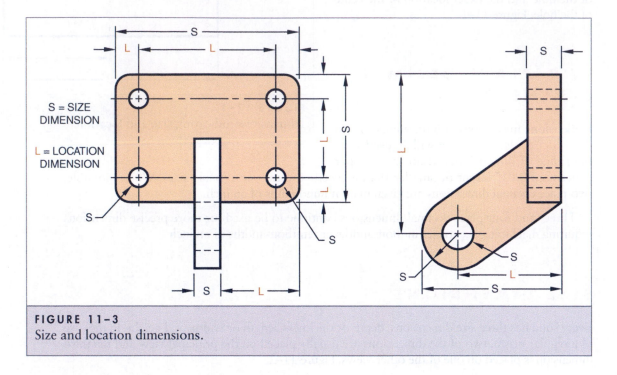

FIGURE 11–3
Size and location dimensions.

PLACING DIMENSIONS

In dimensioning a drawing, the first step is to place extension lines and external center lines where needed, Figure 11–4.

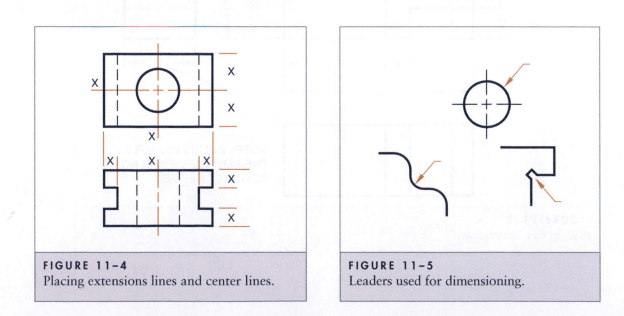

FIGURE 11–4
Placing extensions lines and center lines.

FIGURE 11–5
Leaders used for dimensioning.

Dimension lines and leaders are added next. The term leader refers to a thin, inclined straight line terminating in an arrowhead. The leader directs attention to a dimension or note and the arrowhead identifies the feature to which the dimension or note refers. A few sample leaders are given in Figure 11–5.

The common practice in placing dimensions is to keep them outside the outline of the object. The exception is where the drawing may be made clearer by inserting the dimensions within the object. When a dimension applies to two views, it should be placed between the two views, as shown in Figure 11–6.

Continuous Dimensions

Sets of dimension lines and numerals should be placed on drawings close enough so they may be read easily without any possibility of confusing one dimension with another. If a series of dimensions is required, the dimensions should be placed in a line as continuous dimensions, Figure 11–7A. This method is preferred over the staggering of dimensions, Figure 11–7B, because of ease in reading, appearance, and simplified dimensioning.

Continuous dimensioning is also called chain dimensioning, whereby each dimension is dependent on the previous dimension. Chain dimensioning is commonly used on architectural and construction drawings. It is also used in computer-controlled applications requiring a high degree of dimensional accuracy.

FIGURE 11–6
Placing dimensions between views.

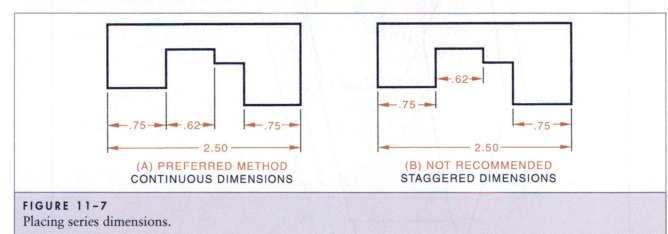

FIGURE 11–7
Placing series dimensions.

DIMENSIONS IN LIMITED SPACES

In the dimensioning of grooves or slots, the dimension, in many cases, extends beyond the width of the extension line. In such instances, the dimension is placed on either side of the extension line or a leader is used, Figure 11–8 (A, B, C, and D). Dots are also used on drawings for dimensioning when space is limited as shown in Figure 11–8E.

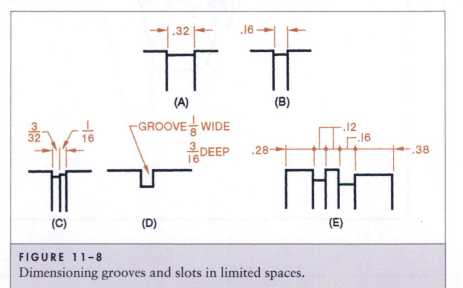

FIGURE 11–8
Dimensioning grooves and slots in limited spaces.

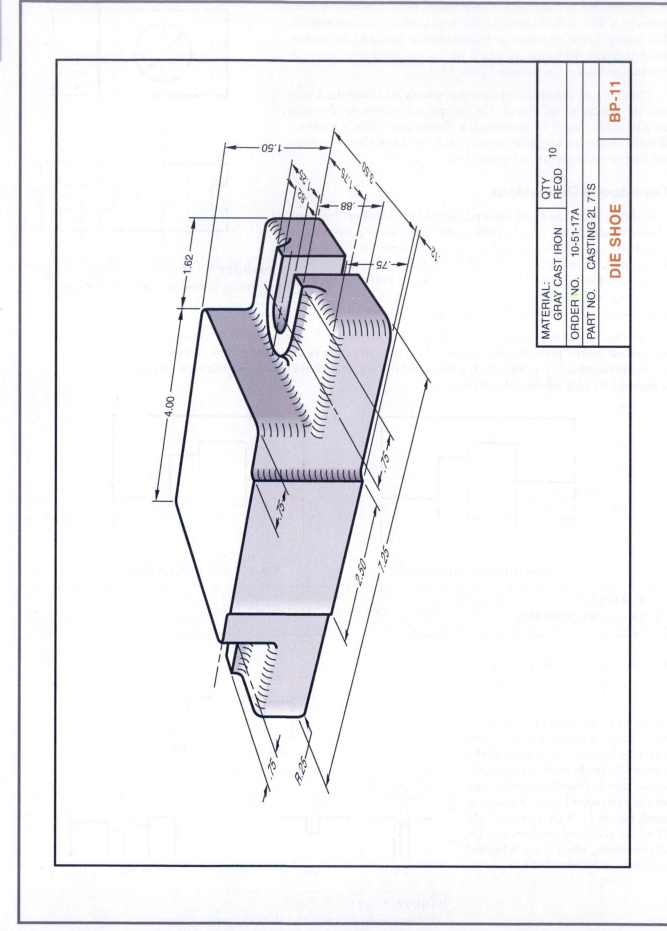

DIE SHOE

MATERIAL: GRAY CAST IRON	QTY REQD	10
ORDER NO. 10-51-17A		
PART NO. CASTING 2L 71S		

BP-11

1.50
1.62
4.00
.82
1.25
1.75
3.50
.88
.75
.12
.75
.75
2.50
7.25
.75
R.25

ASSIGNMENT—UNIT 11: DIE SHOE (BP-11)

Student's Name _____

NOTE: Study the sketch of the Die Shoe.

1. Give the dimensions in both views in the circles provided.

2. Indicate one letter on the two-view drawing that identifies each of the following types of lines:

   (a) Extension _____ ◯

   (b) Dimension _____ ◯

   (c) Object _____ ◯

   (d) Center line _____ ◯

   (e) Hidden edge _____ ◯

3. Place an X in the correct block to identify the kind of dimension to which each letter refers.

Dimension	Size	Location
		Dimension
	Dimension	
G is a	☐	☐
H is a	☐	☐
I is a	☐	☐

12 Dimensioning Cylinders, Circles, and Arcs

ALIGNED AND UNIDIRECTIONAL METHODS OF PLACING AND READING DIMENSIONS

There are two standard methods of placing dimensions. In the older (almost obsolete) method, called the aligned method, each dimension is placed in line with the dimension to which it refers, Figure 12–1A. The second method, the unidirectional method (recommended by ASME), has all numbers or values placed horizontally (one direction), regardless of the direction of the dimension line. All values are read from the bottom, Figure 12–1B.

The aligned and unidirectional methods are illustrated in Figures 12–1A and Figure 12–1B on similar drawings of the same part. Note at (A) that the aligned dimensions are read from both the bottom and the right-side. By contrast, the unidirectional dimensions, Figure 12–1B, are read from the bottom (one direction only).

DIMENSIONING CYLINDERS

The length and diameter of a cylinder are usually placed in the view that shows the cylinder as a rectangle, Figure 12–2. This method of dimensioning is preferred because on small diameter cylinders and holes, a dimension placed in the hole is confusing.

Many round parts, with cylindrical surfaces symmetrical about the axis, can be represented on one-view drawings. The abbreviation for diameter, DIA, is used with the dimension in such instances because no other view is needed to show the shape of the surface, Figure 12–3A. A diameter may also be identified according to ASME standards by using the symbol Ø preceding the dimension. On two-view drawings, DIA may be omitted, Figure 12–3B. In other words, when a cylinder is dimensioned, DIA should follow the dimension unless it is evident that the dimension refers to a diameter.

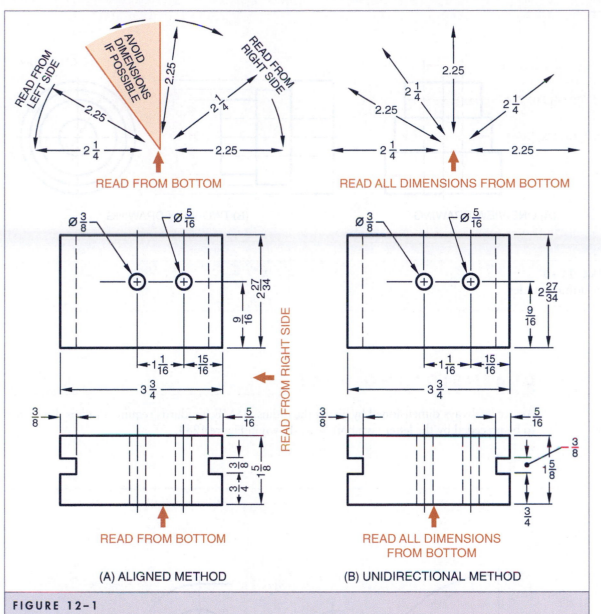

FIGURE 12-1
Aligned and unidirectional dimensioning.

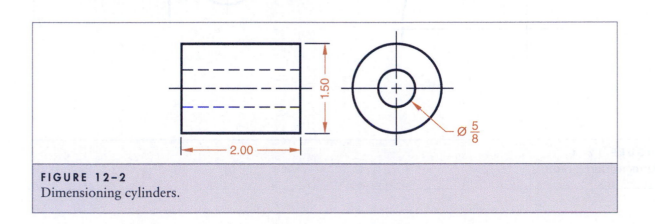

FIGURE 12-2
Dimensioning cylinders.

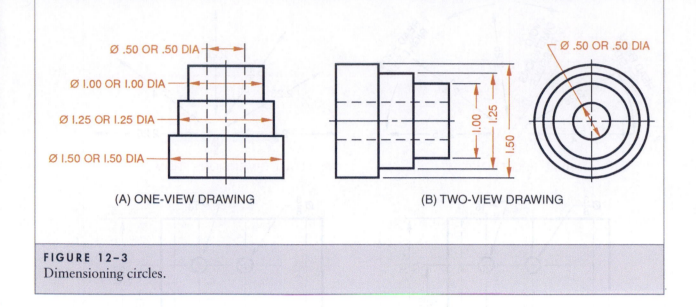

FIGURE 12–3
Dimensioning circles.

DIMENSIONING ARCS

An arc is always dimensioned by giving the radius. ASME standards require a radius dimension to be preceded by the letter (symbol) **R,** as shown in Figure 12–4.

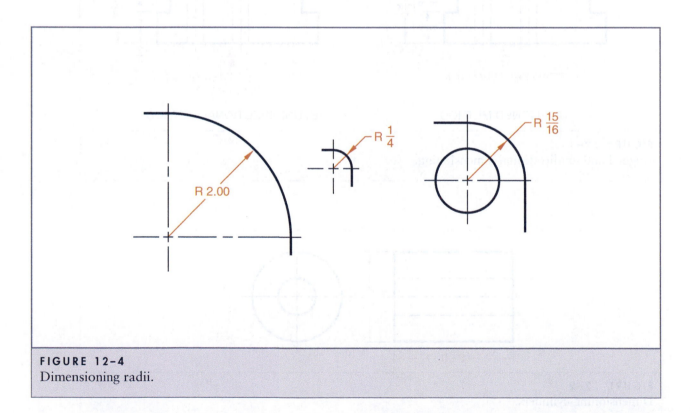

FIGURE 12–4
Dimensioning radii.

DIMENSIONING RADII FROM UNLOCATED CENTERS

In standard practice, center, extension, and dimension lines are usually given on a drawing for locating and dimensioning a radius. Many times, the center is not important, since the location of a radius (arc) is controlled by features or dimensions. Dimensioning in such cases is simplified by conveniently placing the radius dimensions, as shown in Figure 12–5.

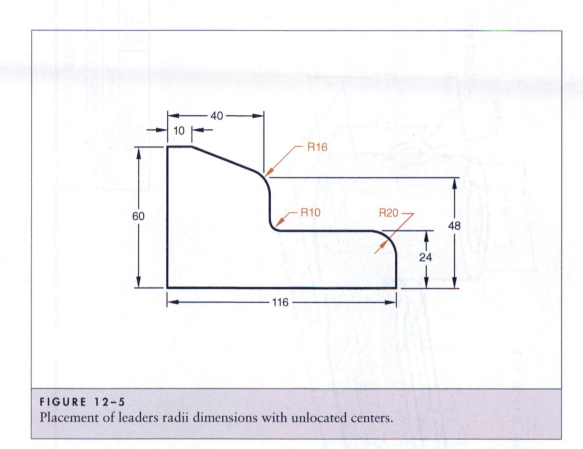

FIGURE 12–5
Placement of leaders radii dimensions with unlocated centers.

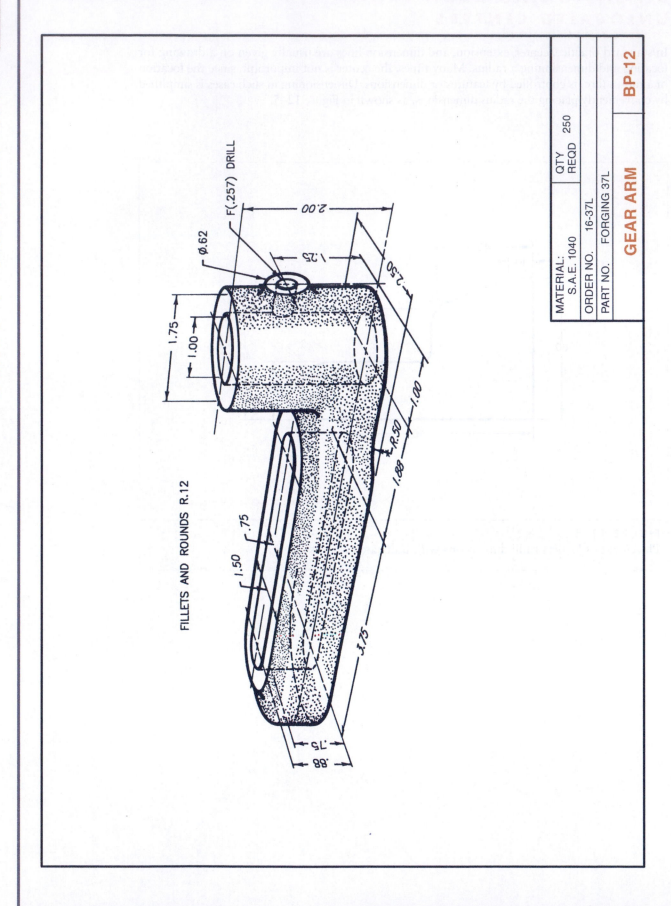

FILLETS AND ROUNDS R.12

F(.257) DRILL

Ø.62

2.00

1.25

2.50

1.75

1.00

R.50

1.88

1.00

.75

1.50

.75

3.75

.75

.88

MATERIAL: S.A.E. 1040		QTY REQD 250
ORDER NO. 16-37L		
PART NO. FORGING 37L		
GEAR ARM		**BP-12**

GEAR ARM (BP-11)

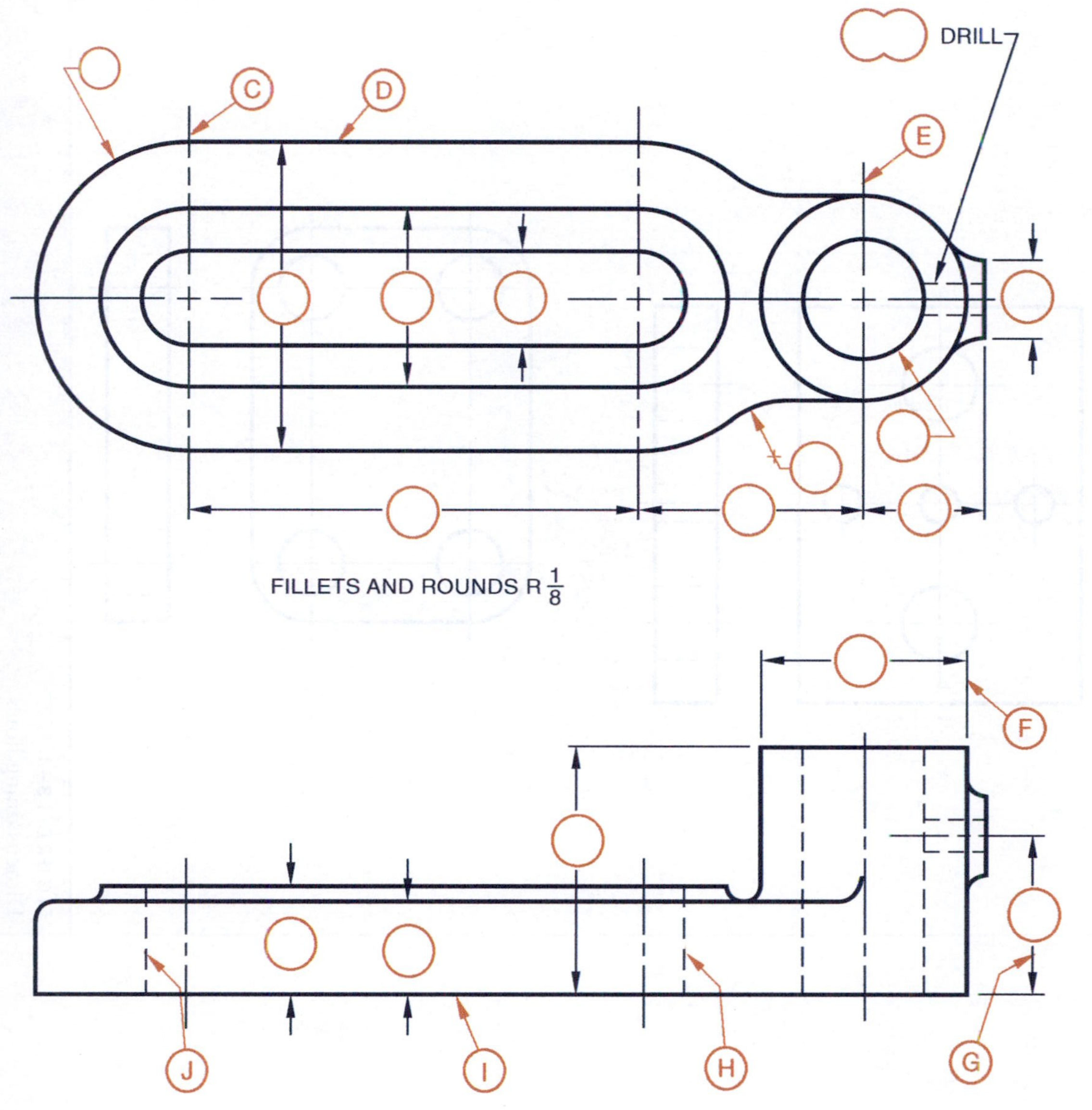

DRILL

FILLETS AND ROUNDS R $\frac{1}{8}$

ASSIGNMENT—UNIT 12: GEAR ARM (BP-12)

Student's Name _____

1. Give the dimensions required in both views in the circles provided, by using the unidirectional method.

2. What is the outside diameter of the upright portion? _____

3. Give the overall length of the elongated slot. _____

4. Determine the overall length of the Gear Arm. _____

5. Name the two views. _____ and _____ views.

6. Refer to the two-view drawing. Give one letter that identifies each type of line.

   (a) Center line _____

   (b) Object line _____

   (c) Extension line _____

   (d) Hidden edge _____

   (e) Dimension line _____

13 | Size Dimensions for Holes and Angles

DIMENSIONING HOLES

Several job shops give diameters of holes that are either formed by drilling, boring, reaming, or punching with a leader, followed by a note like ①, indicating the type of operation to be performed, and ②, indicating the number of holes to be produced, Figure 13–1.

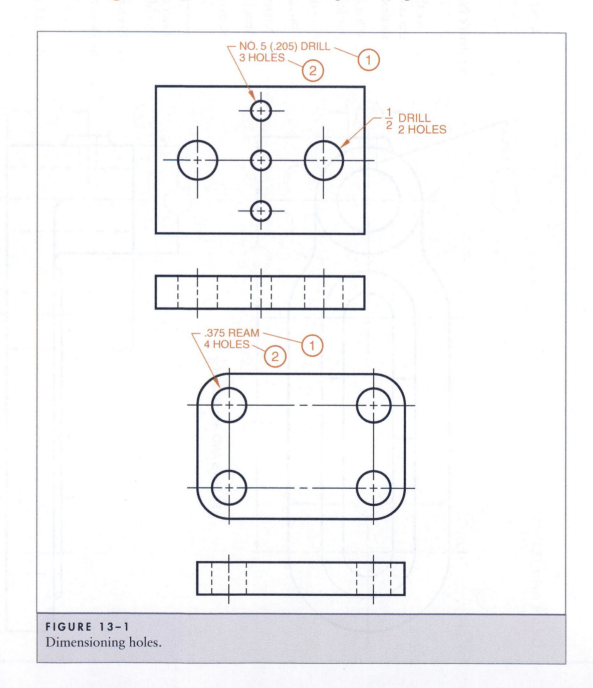

FIGURE 13–1
Dimensioning holes.

DIMENSIONING COUNTERBORED HOLES

A counterbored hole, Figure 13–2, is one that has been machined to a larger diameter for a specified depth so that a bolt or pin will fit into this recessed hole. The counterbored hole provides a flat surface for the bolt or pin to seat against.

Counterbored holes are dimensioned by giving ① the diameter of the drill, ② the diameter of the counterbore, ③ the depth, and ④ the number of holes, Figure 13–2(A). The dimensioning at (B) shows the use of the ASME symbol ⌴ to represent a counterbored hole and ↧ to indicate depth. In the example, the number of holes (2X) is followed by the counterbore symbol ⌴, the diameter symbol Ø, and the diameter 5/8. The depth symbol ↧ is followed by the dimension 5/16.

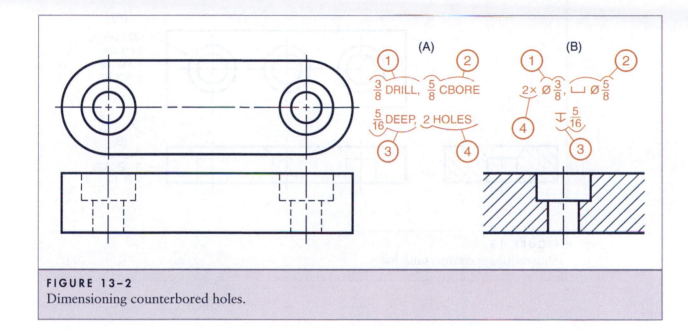

FIGURE 13–2
Dimensioning counterbored holes.

DIMENSIONING COUNTERSUNK HOLES

A countersunk hole, Figure 13–3, is a cone-shaped recess machined in a part to receive a cone-shaped flathead screw or bolt.

Countersunk holes are dimensioned by giving ① the diameter of the hole, ② the angle at which the hole is to be countersunk, ③ the diameter at the large end of the hole, and ④ the number of holes to be countersunk, Figure 13–3A. When the ASME symbol for countersunk V is used, the countersunk holes are dimensioned as shown in Figure 13–3B; ① number of holes 3X; ② diametral symbol and diameter (Ø .328); ③ countersunk hole symbol, V, diametral symbol (Ø), and diameter (.62); and the included angle of the countersink (82°).

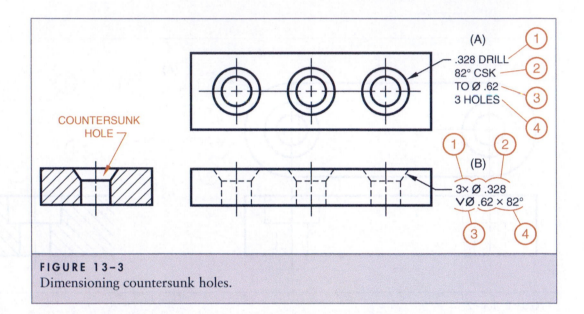

FIGURE 13-3
Dimensioning countersunk holes.

DIMENSIONING ANGLES

The design of a part may require some lines to be drawn at an angle. The amount of the divergence (the amount the lines move away from each other) is indicated by an angle measured in degrees or fractional parts of a degree. The degree is indicated by the symbol ° placed after the numerical value of the angle. For example, in Figure 13–4B, 45° indicates that the angle measures 45 degrees. Two common methods of dimensioning angles show (1) linear dimensions or (2) the angular measurement, as illustrated in Figure 13–4.

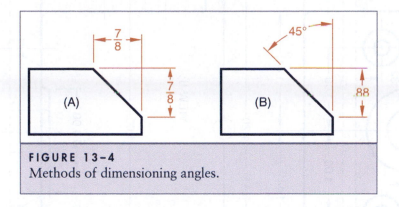

FIGURE 13–4
Methods of dimensioning angles.

The dimension line for an angle should be an arc whose ends terminate in arrowheads. The numeral indicating the degrees in the angle is read in a horizontal position, except where the angle is large enough to permit the numerals to be placed along the arc, Figure 13–5.

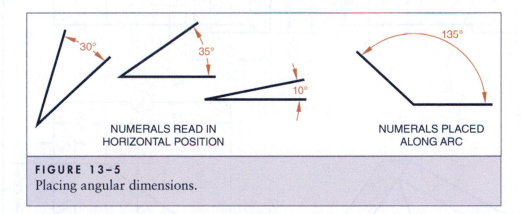

FIGURE 13–5
Placing angular dimensions.

DIMENSIONING CHORDS AND ARCS

Dimensioning of chords and arcs according to ASME standards is illustrated in Figure 13–6. The symbol ⌒ is used to indicated an arc.

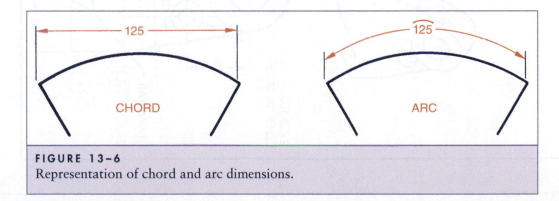

FIGURE 13–6
Representation of chord and arc dimensions.

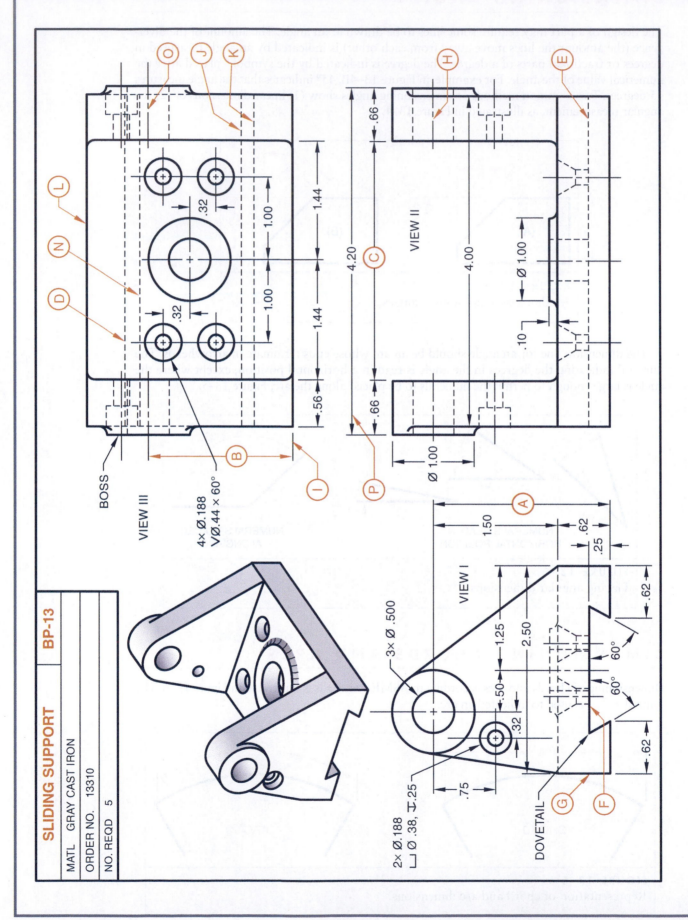

SLIDING SUPPORT   BP-13

MATL   GRAY CAST IRON

ORDER NO.   13310

NO. REQD   5

ASSIGNMENT—UNIT 13: SLIDING SUPPORT (BP-13)

Student's Name _____

1. Name VIEW I, which shows the shape of the dovetail.

2. Name VIEW III in which the base pad appears as a circle.

3. Name VIEW II.

4. Name the kind of line shown at Ⓔ.

5. What surface in VIEW I is represented by line Ⓔ?

6. Name the kind of line shown at Ⓖ.

7. What line in VIEW III represents surfaces Ⓖ?

8. Name the kind of line shown at Ⓗ.

9. What line in VIEW III represents the line Ⓗ?

10. Name the kind of line shown at Ⓘ.

11. Name the kind of line shown at Ⓙ.

12. What lines in the top view represent the dovetail?

13. What does the line Ⓔ in the front view represent?

14. Determine height Ⓐ.

15. How many bosses are shown on the uprights?

16. What is the outside diameter of the boss?

17. Determine dimension Ⓑ.

18. How far off from the center of the support is the center of the two holes in the bosses of the uprights?

19. Give the dimensions for the counterbored holes.

20. What dimensions are given for the countersunk holes?

21. Give the diameter of the Ⓑ largest holes.

22. What is the basic dimension Ⓒ?

23. How wide is the opening in the dovetail?

24. How deep is the dovetail machined?

25. What is the angle to the horizontal at which the dovetail is cut?

1. _____
2. _____
3. _____
4. _____
5. _____
6. _____
7. _____
8. _____
9. _____
10. _____
11. _____
12. _____
13. _____
14. _____
15. _____
16. _____
17. _____
18. _____
19. _____
20. _____
21. _____
22. _____
23. _____
24. _____
25. _____

14. Location Dimensions for Points, Centers, and Holes

DIMENSIONING A POINT OR A CENTER

A point or a center of an arc or circle is generally measured from two finished surfaces. This method of locating the center is preferred to making an angular measurement.

In Figure 14–1, the center of the circle and arc may be easily found by measuring the vertical and horizontal center lines from the machined surfaces.

DIMENSIONING EQUALLY SPACED HOLES ON A CIRCLE

If a number of holes are to be equally spaced on a circle, the exact location of the first hole is given by location dimensions. To locate the remaining holes, the location dimensions are followed by ① the diameter of the holes, ② the number of holes, and ③ the notation equally spaced or eq sp, Figure 14–2. If

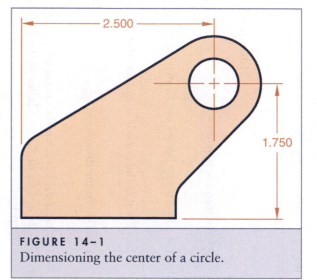

FIGURE 14–1
Dimensioning the center of a circle.

following ASME standards, Figure 14–2B will then show the same information but in a different sequence and will also use symbols as noted using the color BLUE. Note that in Figure 14–2B, the notation "eq sp" is not used. It is replaced by 4 × 45° at ③.

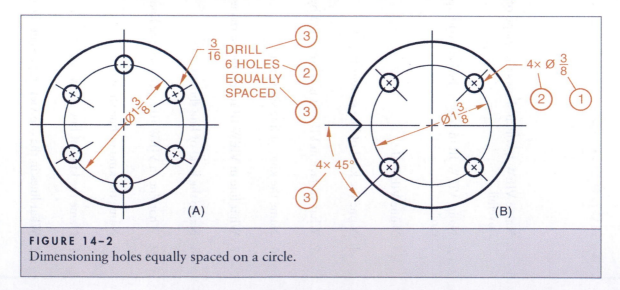

FIGURE 14–2
Dimensioning holes equally spaced on a circle.

DIMENSIONING UNEQUALLY SPACED HOLES ON A CIRCLE

When holes are to be located on a circle, the diameter of the circle should be given so as to fix the exact center of each hole. The size and position of each hole are noted on the drawing, Figure 14–3A. If more than one hole is the same diameter, then a notation may be used to indicate this fact, Figure 14–3B.

For example, the notation 5X Ø.500 means that the five holes on the drawing are 1/2″ in diameter.

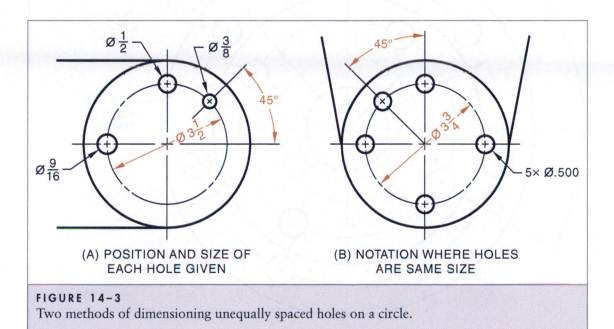

(A) POSITION AND SIZE OF EACH HOLE GIVEN

(B) NOTATION WHERE HOLES ARE SAME SIZE

FIGURE 14–3
Two methods of dimensioning unequally spaced holes on a circle.

DIMENSIONING HOLES NOT ON A CIRCLE

Holes are often dimensioned in relation to one another and to a finished surface. Dimensions are usually given, in such cases, in the view that shows the shape of the holes; that is, square, round, or elongated. The preferred method of placing these dimensions is shown in Figure 14–4A.

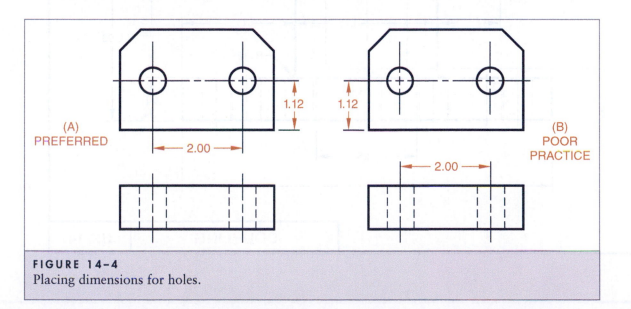

(A) PREFERRED

(B) POOR PRACTICE

FIGURE 14–4
Placing dimensions for holes.

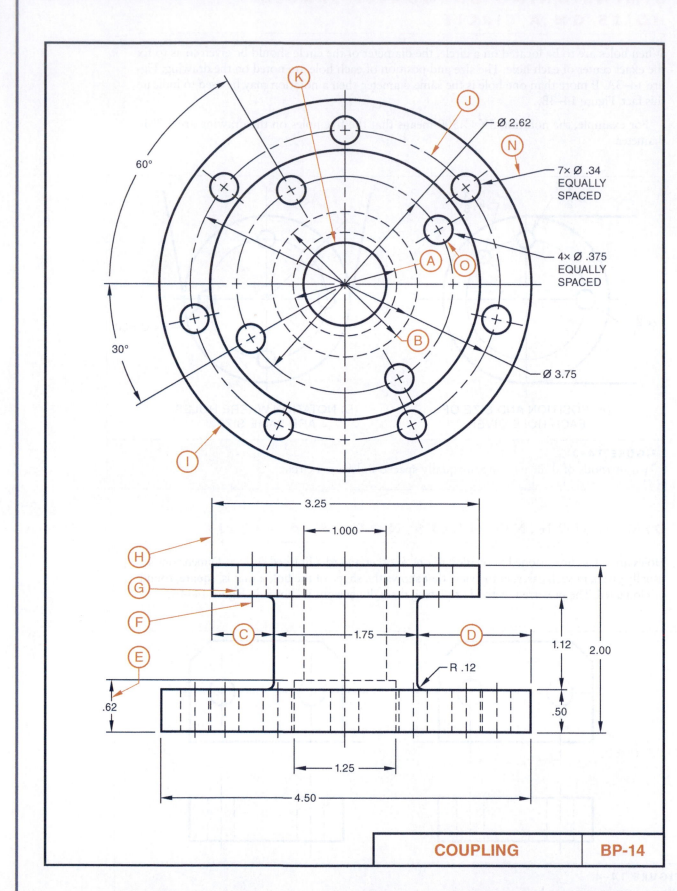

When holes are located on a circle, the diameter of the circle should be given as is the size of the circle of each hole. The size and location of each hole are noted on the drawing. If more than one hole is the same diameter, that a note is given. Use a leader to link up. (See also Figure 18-3B.)

For example, 7× means there are seven similar holes on this center circle, not on this figure.

COUPLING   **BP-14**

ASSIGNMENT—UNIT 14: COUPLING (BP-14)

Student's Name _____

1. Name the view that shows the width (height) of the coupling.

2. Name the view in which the holes are shown as circles.

3. Name the kind of line shown at Ⓔ, Ⓕ, Ⓖ, Ⓗ, Ⓘ, and Ⓙ.

4. What circle represents the 4.50″ diameter flange?

5. What circle represents the 1.000″ diameter?

6. Name the kind of line shown at Ⓝ.

7. How many holes are to be drilled in the larger flange?

8. Indicate the drill size to be used.

9. Give the diameter circle on which the equally spaced holes are drilled in the larger flange.

10. How many holes are to be drilled in the smaller flange?

11. How deep is the 1.25″ diameter hole?

12. Give the diameter of the reamed holes.

13. State the angle with the horizontal center line used for locating reamed hole Ⓞ.

14. What is the overall width (height) of the coupling?

15. What is the diameter of the smaller flange?

16. What is the diameter of the circle Ⓐ and Ⓑ?

17. What is the depth of the 1.000″ hole?

18. What is the thickness of the larger flange?

19. If 5/16″ bolts are used in the 7 drilled holes, what will be the clearance between the hole and the bolt?

20. Determine distances Ⓒ and Ⓓ.

1. _____

2. _____

3. Ⓔ _____

   Ⓕ _____

   Ⓖ _____

   Ⓗ _____

   Ⓘ _____

   Ⓙ _____

4. _____

5. _____

6. _____

7. _____

8. _____

9. _____

10. _____

11. _____

12. _____

13. _____

14. _____

15. _____

16. Ⓐ _____

    Ⓑ _____

17. _____

18. _____

19. _____

20. Ⓒ _____

    Ⓓ _____

15 Dimensioning Large Arcs and Base Line Dimensions

DIMENSIONING ARCS WITH CENTERS OUTSIDE THE DRAWING

When the center of an arc falls outside the limits of the drawing, a broken dimension line is used, as illustrated in Figure 15–1. This dimension line gives the size of the arc and indicates that the arc center lies on a center line outside the drawing. This technique is also used when a dimension line interferes with other parts of a drawing.

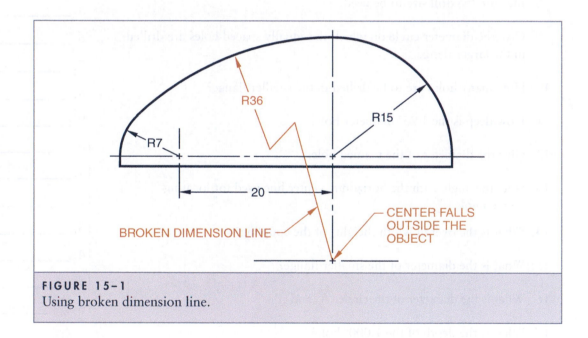

FIGURE 15–1
Using broken dimension line.

BASE LINE DIMENSIONING

In base line dimensioning, all measurements are made from common finished surfaces called base lines or reference lines, Figure 15–2. Base line dimensioning is used when accurate layout work to precision limits is required. Errors are not cumulative with this type of dimensioning because all measurements are taken from the base lines.

Dimensions and measurements may be taken from one or more base lines. In Figure 15–2, the two base lines are at right angles to each other. The horizontal dimensions are measured from base line (B), which is a machined edge. The vertical dimensions are measured from surface (A), which is at a right angle to surface (B) and is also a machined surface.

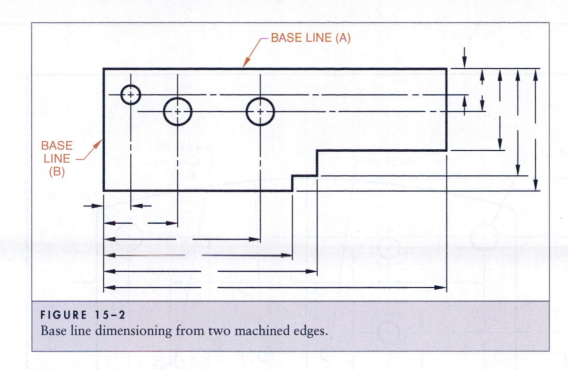

FIGURE 15–2
Base line dimensioning from two machined edges.

An application of base line dimensioning, where a center line is used as the reference line, is shown in Figure 15–3A. Base line dimensioning may also be applied to irregular shapes, such as the template shown in Figure 15–3B.

Base line dimensions simplify the reading of a drawing and also permit greater accuracy in making the part.

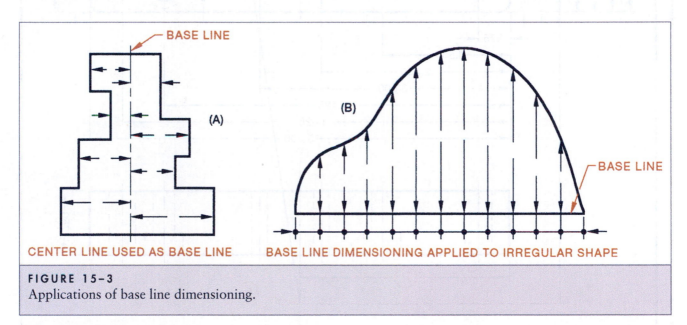

FIGURE 15–3
Applications of base line dimensioning.

STATING INCHES

Dimensions under 72 inches are usually stated in inches, so the inch (") symbol may be omitted on a drawing. This applies primarily to manufacturing industries. By comparison, measurements for structural work and the building industry are usually given in feet and inches.

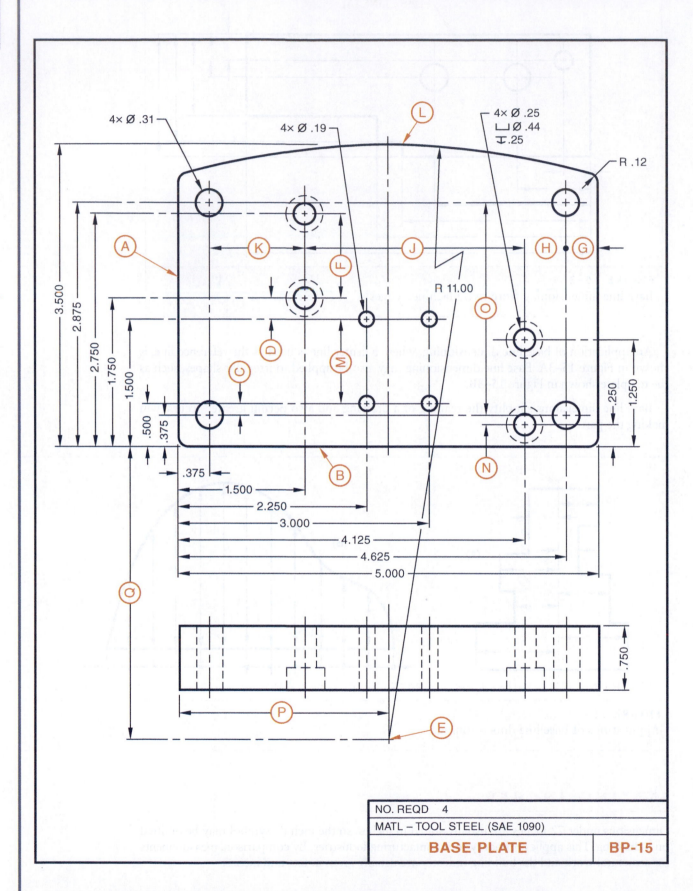

4× Ø .31
4× Ø .19
L
4× Ø .25
⌴ Ø .44
⌁ .25
R .12

A
K
J
H G
F
D
M
C
O
B
N

R 11.00

3.500
2.875
2.750
1.750
1.500
.500
.375

.375
1.500
2.250
3.000
4.125
4.625
5.000

Q

.250
1.250

.750

P

E

NO. REQD	4
MATL – TOOL STEEL (SAE 1090)	
BASE PLATE	**BP-15**

ASSIGNMENT—UNIT 15: BASE PLATE (BP-15)

Student's Name _____

1. Give the name of the part.

2. What material is used for the base part?

3. How many parts are required?

4. What is the length of the base plate?

5. What is the height of the base plate?

6. What is the thickness (depth) of the base plate?

7. How many .31″ holes are to be drilled?

8. How many .19″ holes are to be drilled?

9. How many .25″ holes are to be drilled?

10. Give (a) the diameter and (b) the depth of counterbore for the .25″ holes.

11. What system of dimensioning is used on this drawing?

12. Give the letter of the base line in the front view from which all depth dimensions are taken.

13. Give the letter of the base line in the front view from which all horizontal dimensions are taken.

14. Compute the following height dimensions: Ⓒ, Ⓓ, Ⓜ, and Ⓕ.

15. Compute the following vertical dimensions: Ⓖ, Ⓗ, Ⓙ, and Ⓚ.

16. Compute dimensions Ⓝ and Ⓞ.

17. Give the radius to which the corners are rounded.

18. What is the radius of arc Ⓛ?

19. What letter indicates the center for arc Ⓛ?

20. Compute dimensions Ⓟ and Ⓠ.

1. _____

2. _____

3. _____

4. _____

5. _____

6. _____

7. _____

8. _____

9. _____

10. (a) _____ (b) _____

11. _____

12. _____

13. _____

14. Ⓒ = _____

Ⓓ = _____

Ⓜ = _____

Ⓕ = _____

15. Ⓖ = _____

Ⓗ = _____

Ⓙ = _____

Ⓚ = _____

16. Ⓝ = _____

Ⓞ = _____

17. _____

18. _____

19. _____

20. Ⓟ = _____

Ⓠ = _____

16

Tolerances: Fractional and Angular Dimensions

TOLERANCES

As a part is planned, the designer must consider (1) its function either as a separate unit or as a part that must move in a fixed position in relation to other parts, (2) the operations required to produce the part, (3) the material to be used, (4) the quantity to be produced, and (5) the cost. Each of these factors influences the degree of accuracy to which a part is machined.

The dimensions given on a drawing are an indication of what the limits of accuracy are. These limits are called tolerances. Parts may have a tolerance given in fractions or decimal inches or decimal millimeters. CAD drawings can dimension in any of these systems. For CNC-manufactured parts, decimal dimensions are required.

SPECIFYING FRACTIONAL TOLERANCES

The note in Figure 16–1, TOLERANCES ON FRACTIONAL DIMENSIONS ARE ± 1/64″, indicates that the dimension given in fractions on the drawing may be any size between a 64th of an inch larger to a 64th of an inch smaller than the specified size.

For example, on the 2 1/2″ dimension in Figure 16–1:

1. The tolerance given on the drawing is ±1/64″.
2. The largest size to which the part may be machined is 2 1/2″ + 1/64″ = 2 33/64″.
3. The smallest size to which the part may be machined is 2 1/2″ − 1/64″ = 2 31/64″.

The larger size is called the upper-limit; the smaller size is called the lower-limit.

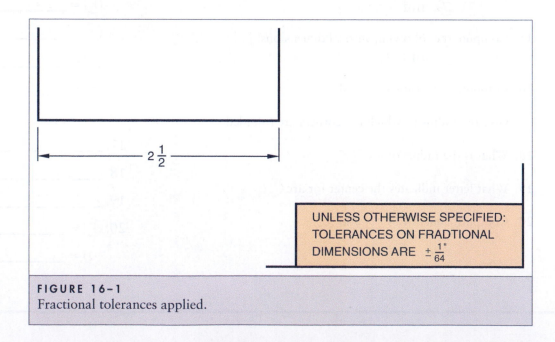

FIGURE 16–1
Fractional tolerances applied.

ANGULAR DIMENSIONS

Angles may be dimensioned in degrees or parts of a degree.

1. Each degree is one three-hundred-sixtieth of a circle (1/360).
2. The degree may be divided into smaller units called **minutes.** There are 60 minutes in each degree.
3. Each minute may be divided into smaller units called **seconds.** There are 60 seconds in each minute.

To simplify the dimensioning of angles, symbols are used to indicate degrees, minutes, and seconds, Figure 16–2. For example, twelve degrees, fifteen minutes, and forty-five seconds can also be written 12° 15′ 45″.

Decimalized angles are now preferred. To convert angles given in whole degrees, minutes, and seconds, the following steps should be followed.

Example: Convert 12° 15′ 45″ into decimal degrees.

1. Convert minutes into degrees by dividing by 60 (60′ = 1°).

    15 ÷ 60 = .25°

2. Convert seconds into degrees by dividing seconds by 3600 (3600″ = 1°).

    45″ ÷ 3600 = .01°

3. Add whole degrees plus decimal degrees.

    12° + 25° + .01° = 12.26°

    therefore, 12° 15′ 45″ = 12.26° decimal degrees.

	Symbol
Degrees	°
Minutes	′
Seconds	″

FIGURE 16–2
Symbols used for dimensioning angles.

SPECIFYING ANGULAR TOLERANCES

The tolerance on an angular dimension may be given in a note on the drawing, as shown in Figure 16–3. The tolerance may also be shown on the angular dimension itself, Figure 16–4.

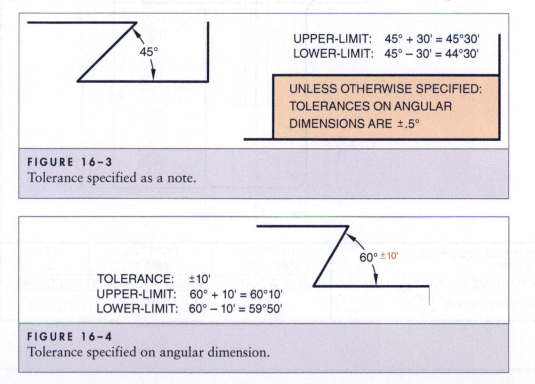

45°

UPPER-LIMIT: 45° + 30′ = 45°30′
LOWER-LIMIT: 45° − 30′ = 44°30′

UNLESS OTHERWISE SPECIFIED:
TOLERANCES ON ANGULAR
DIMENSIONS ARE ±.5°

FIGURE 16–3
Tolerance specified as a note.

TOLERANCE: ±10′
UPPER-LIMIT: 60° + 10′ = 60°10′
LOWER-LIMIT: 60° − 10′ = 59°50′

60° ±10′

FIGURE 16–4
Tolerance specified on angular dimension.

UNIT 16 | TOLERANCES: FRACTIONAL AND ANGULAR DIMENSIONS

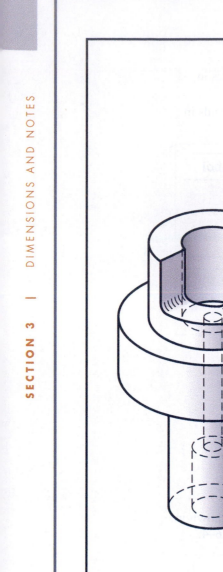

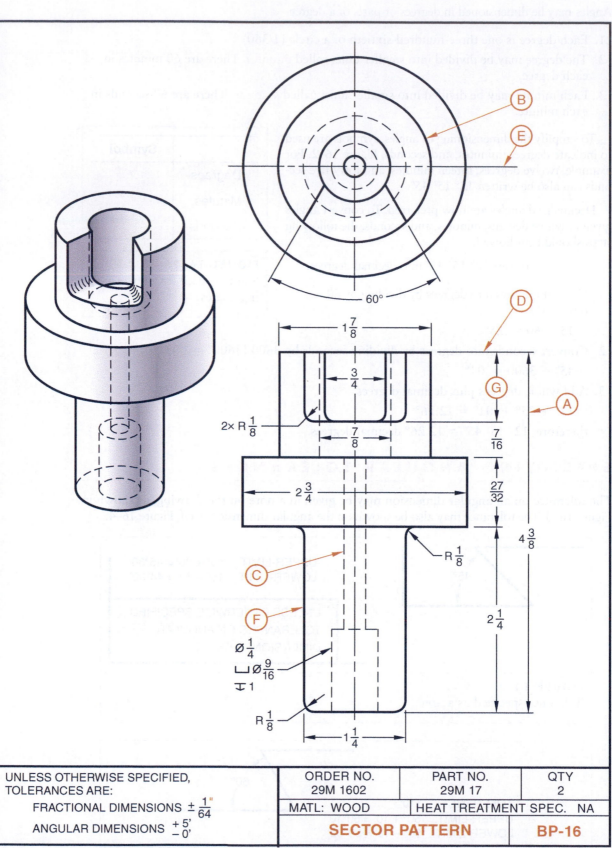

60°

1 7/8

3/4

2× R 1/8

7/8

2 3/4

7/16

27/32

4 3/8

2 1/4

R 1/8

Ø 1/4

Ø 9/16

1

R 1/8

1 1/4

UNLESS OTHERWISE SPECIFIED, TOLERANCES ARE:	ORDER NO. 29M 1602	PART NO. 29M 17	QTY 2
FRACTIONAL DIMENSIONS ± 1/64"	MATL: WOOD	HEAT TREATMENT SPEC. NA	
ANGULAR DIMENSIONS + 5' − 0'	**SECTOR PATTERN**		**BP-16**

ASSIGNMENT—UNIT 16: SECTOR PUNCH (BP-16)

Student's Name _____

1. What letter is used to denote a

    (a) hidden edge line?   1. (a) _____

    (b) dimension line?   (b) _____

    (c) extension line?   (c) _____

    (d) center line?   (d) _____

2. Determine basic dimension (G).   2. _____

3. What is the diameter of the drilled pilot hole for the   3. _____
   counterbore?

4. Give the diameter and depth (heighth) of the   4. Diameter _____
   counterbore.
                                                              Depth _____

5. What are the upper- and lower-limits of tolerance for   5. _____
   fractional dimensions?
                                                              _____

6. What limit of tolerance is specified for angular   6. _____
   dimensions?

7. What is the largest size to which the 2 3/4″ diameter   7. _____
   can be turned?

8. What is the lower-limit to which the 2 3/4″ diameter   8. _____
   can be machined?

9. Give the upper-limit to which diameter (B) may be   9. _____
   machined.

10. Give the upper- and lower-limit on the diameter for   10. Upper _____
    shank (F).
                                                              Lower _____

11. If shank (F) is machined to the upper-limit length   11. _____
    (height), how long will it be?

12. If shank (F) is machined 2 3/16″ long, how much   12. _____
    under the lower-limit size will it be?

13. How much over the upper-limit will the shank be if it 13. _____
    is 2 5/16″ long?

14. What is the upper-limit of accuracy for the 60° angle? 14. _____

15. If the height of the 1 7/8″ diameter punch measures   15. _____
    1 9/32″, is it over, under, or within the specified limits
    of accuracy?

17 Tolerances: Unilateral and Bilateral, Decimal Dimensions

UNILATERAL AND BILATERAL TOLERANCES

The limits of accuracy to which a part is to be produced may fall within one of two classifications of tolerances. A dimension is said to have a **unilateral (single) tolerance** when the total tolerance is in one direction only, either (+) or (−). The examples of unilateral tolerances shown in Figure 17–1 indicate that part (A) meets standards of accuracy when the basic dimension varies in one direction only and is between 3″ and 3 1/64″; part (B) may vary from 2.44″ to 2.43″; and angle (C) may vary between 60° and 59.75′.

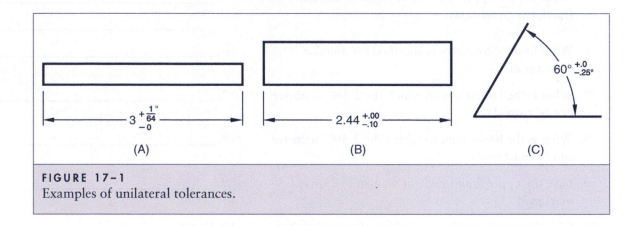

FIGURE 17–1
Examples of unilateral tolerances.

Bilateral tolerances applied to dimensions mean that the dimensions may vary from a larger size (+) to a smaller size (−) than the basic dimension (nominal size). In other words, the basic dimension may vary in both directions. The basic 3″ dimension in Figure 17–2A with a ±1/64″ tolerance may vary between 3 1/64″ and 2 63/64″. The basic 2.44″ dimension in Figure 17–2B with a bilateral tolerance of ±.01″ is acceptable within a range of 2.45″ and 2.43″. The 60° angle in Figure 17–2C with a tolerance of ±.25° may range between 60.25° and 59.75°. These bilateral tolerances are shown in Figure 17–2.

When the dimensions within the tolerance limits appear on a drawing, they are expressed as a range from the smaller to the larger dimension, as indicated in Figure 17–2.

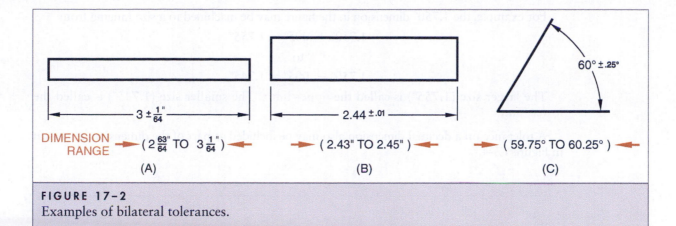

FIGURE 17–2
Examples of bilateral tolerances.

DECIMAL DIMENSIONS

The decimal system of dimensioning is used almost exclusively in industry because of the ease with which computations can be made. In addition, the dimension can be measured with precision instruments to a high degree of accuracy and is required for Computer Numerical Control (CNC) applications.

Dimensions in the decimal system can be read quickly and accurately. Dimensions are read in thousandths, $1/1000'' = (.001'')$; in ten thousandths, $1/10,000'' = (.0001'')$; and in even finer divisions if necessary.

SPECIFYING DECIMAL TOLERANCES

Decimal tolerances can be applied in both the English and the metric systems of measurement. Tolerances on decimal dimensions that are expressed in terms of two, three, four, or more decimal places may be given on a drawing in several ways. One of the common methods of specifying a tolerance that applies on all dimensions is to use a note, Figure 17–3.

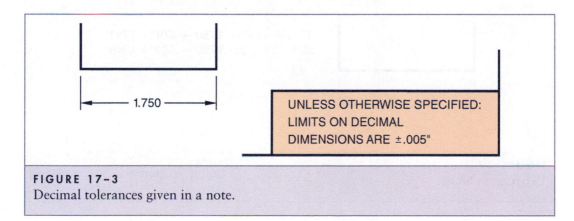

FIGURE 17–3
Decimal tolerances given in a note.

For example, the 1.750″ dimension in the figure may be machined to a size ranging from

$$1.750″ + .005″ = 1.755″$$

to

$$1.750″ − .005″ = 1.745″$$

The larger size (1.755″) is called the **upper-limit.** The smaller size (1.745″) is called the **lower-limit.**

A tolerance on a decimal dimension also may be included as part of the dimension, as shown in Figure 17–4.

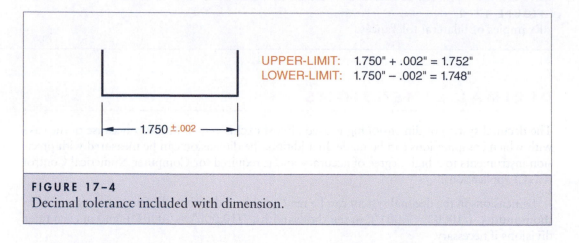

FIGURE 17–4
Decimal tolerance included with dimension.

Bilateral tolerances are not always equal in both directions. It is common practice for a drawing to include either a (+) tolerance or a (−) tolerance that is greater than the other.

In cases where the plus and minus tolerances are not the same, such as plus .001″ and minus .002″, the dimension may be shown on the drawing, as in Figure 17–5.

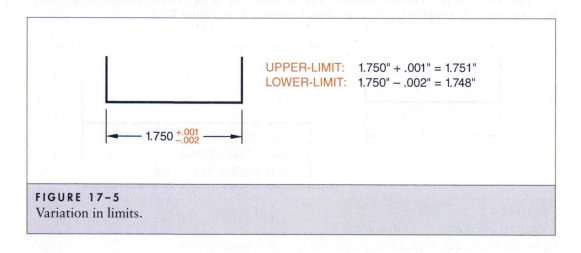

FIGURE 17–5
Variation in limits.

The same variation between the upper- and lower-limit can be given as in Figure 17–6. The dimension above the line is the upper-limit; the dimension below the line is the lower-limit.

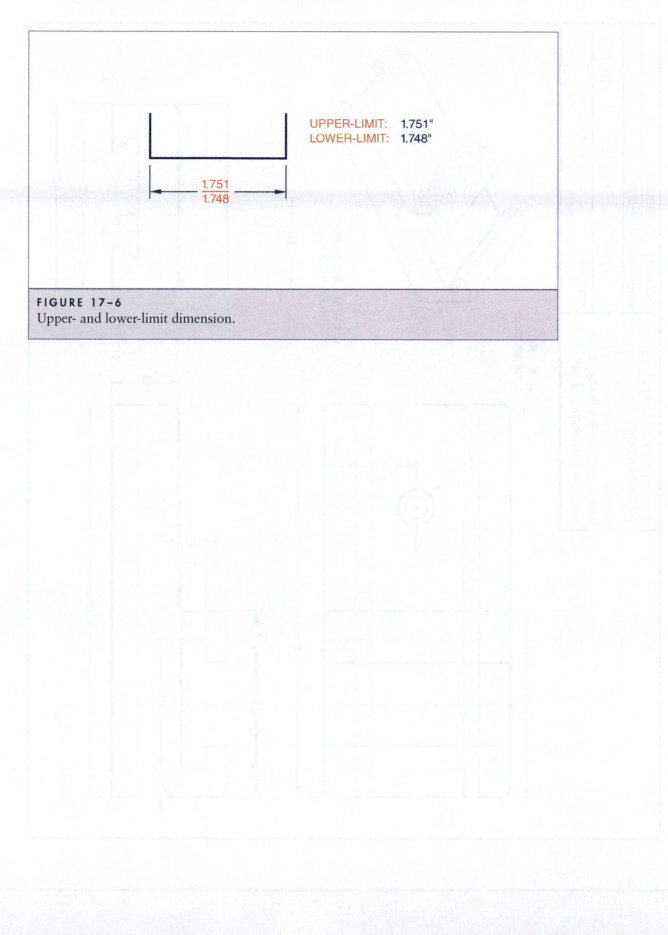

FIGURE 17–6
Upper- and lower-limit dimension.

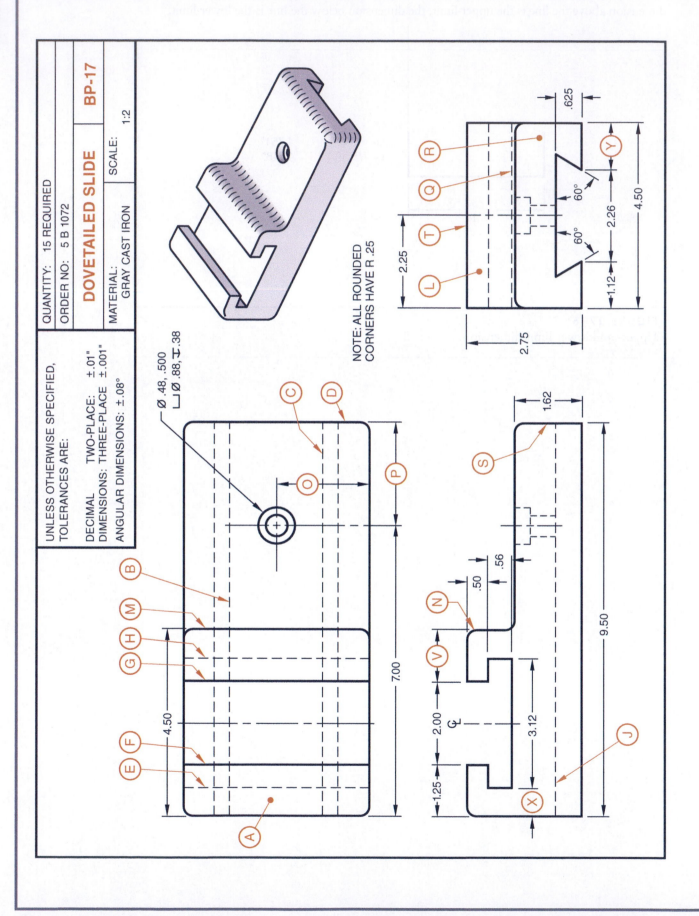

QUANTITY: 15 REQUIRED

ORDER NO: 5 B 1072

DOVETAILED SLIDE   **BP-17**

MATERIAL:
GRAY CAST IRON

SCALE: 1:2

UNLESS OTHERWISE SPECIFIED,
TOLERANCES ARE:

DECIMAL   TWO-PLACE: ±.01"
DIMENSIONS: THREE-PLACE ±.001"
ANGULAR DIMENSIONS: ±.08°

Ø .48, .500
⌴Ø .88, ⊼ .38

NOTE: ALL ROUNDED
CORNERS HAVE R .25

ASSIGNMENT—UNIT 17: DOVETAILED SLIDE (BP-17)

Student's Name _____

1. What type of lines are (B), (C), (H), (E), and (J)?

2. What tolerance is allowed on

   (a) decimal dimensions?

   (b) angular dimensions?

3. What is the minimum overall height of the dovetailed slide?

4. Give the upper-limit dimension for the 60° angle.

5. Give the upper- and lower-limit dimensions for (P).

6. What is the maximum depth (heighth) to which the counterbored hole can be bored?

7. What line in the top view represents surface (R) of the side view?

8. What line in the front view represents surface (L)?

9. What line in the side view represents surface (A) of the top view?

10. What dimension indicates how far line (J) is from the base of the slide?

11. What two lines in the top view indicate the opening of the dovetail?

12. How wide is the opening in the dovetail?

13. At what angle to the horizontal is the dovetail cut?

14. Give dimension (Y).

15. To what depth into the piece is the dovetail cut?

16. What is the horizontal distance from surface (Q) to surface (T)?

17. What is the upper-limit dimension between surfaces (F) and (G)?

18. What is the full depth of the tee slot?

19. Compute dimensions (V) and (X).

20. What is the vertical distance from line (N) to line (S)?

21. What classification of tolerances applies to all dimensions?

22. Change the decimal and angular tolerances so that only the (−) tolerances apply. Then, determine the upper- and lower-limit dimensions for:

    (a) The angular dimension.

    (b) The distance between surfaces (F) and (G).

    (c) Distance (P).

1. _____
2. (a) _____
   (b) _____
3. _____
4. _____
5. _____
6. _____
7. _____
8. _____
9. _____
10. _____
11. _____
12. _____
13. _____
14. _____
15. _____
16. _____
17. _____
18. _____
19. _____
20. _____
22. (a) _____
    (b) _____
    (c) _____

18 Interchangeable Parts, Allowances, and Classes of Fit

Up to this point in the text, we have been dealing with detail drawings, which contain a significant number of views, dimensions, notes, and other information to produce a single part. However, most objects are made of several parts assembled together. Many of these assembled parts will be mass produced.

An essential element of mass production is interchangeable manufacturing, whereby parts can be made in widely separated locations and then brought together for assembly. It makes modern industry possible, just as effective size control allows interchangeable manufacturing to be accomplished.

Unfortunately, it is not possible to make anything to exact size. Parts can be manufactured with a very high degree of accuracy (to within a few millionths of an inch or thousandths of a millimeter), but it is extremely expensive. Fortunately, varying degrees of accuracy determined by functional requirements are all that is needed, not exact sizes.

This means that a way of specifying dimensions with whatever degree of accuracy may be required is necessary. This is accomplished through specifications of a tolerance on each dimension, limits on each part, and fits between mating parts. If allowance and tolerances are properly determined, mating parts can frequently be completely interchangeable. Needing a higher degree of accuracy may require geometric dimension and tolerancing techniques covered in Unit 29.

DESIGN SPECIFICATIONS AND CLASSES OF FITS

Drawings of interchangeable parts contain design data that deal with fits. The term fit is produced by an allowance for clearance, interference, and either clearance or interference in what is know as a transition fit. A drawing dimensioned for a fit provides information about the range of tightness between two mating parts. When the parts move in relation to each other, the dimensions must provide for a positive clearance. The amount of positive clearance depends on the kind of material in the parts, the nature of the motion, lubrication, temperature, and other forces and factors. Machined parts with a positive clearance are dimensioned for a clearance fit. A negative clearance is required with parts that are to be forced together as a single unit. Parts requiring a negative clearance are dimensioned for an interference fit.

TERMS USED WITH ALLOWANCES AND TOLERANCES FOR DIFFERENT CLASSES OF FITS

There are three general classes of fits in both ANSI and ISO metric standards: clearance, transition, and interference. Inch fits are further broken down into five classifications, while there are nine metric.

The following is a summary of common terms used to classify and dimension fits:

Actual Size. True measurement of a part after it had been produced.

Allowance. Prescribed, intentional difference in the dimensions of mating parts. The result of taking the LL hole size minus the UL of its mating shaft hole.

Basic Size. Dimension providing a theoretically exact size, form, or position of a surface, point, or fixture. Provides a size from which limits are derived.

Clearance Maximum. Loosest fit between two mating parts.

Designed Size. Basic size of a feature with tolerances applied.

Fit. Range of tightness or looseness resulting from allowances and tolerances in mating parts.

Clearance Fit. Clearance always exists between mating parts under all tolerance conditions.

Interference Fit. No clearance exists between mating parts under all tolerance conditions.

Transition Fit. Depends on the range of limits in which a clearance or interference can exist.

Limits. Dimensions related to the largest (UL) and smallest (LL) boundary or location acceptable size of a feature.

Nominal Size. General dimension used to identify a commercial product.

Tolerance. Total amount a part can vary from the basic size and still be usable.

AN EXAMPLE OF APPLYING THE ABOVE TERMS TO A MATING SHAFT AND HOLE

Size Terms	Shaft	Hole
Nominal Size	2½	2½
Basic Size	2.500	2.500
Actual Size	2.498	2.500
Designed Size	2.498	2.502
Tolerance	+.000	+.002
	−.002	−.002
Upper-Limit	2.498	2.504
Lower-Limit	2.496	2.500

(1) Max. Clearance = UL Hole   2.504
      − LL shaft   2.496
            +.008

(2) Min. Clearance = UL Hole   2.500
     (allowance) − UL Shaft   2.498
            +.002

Because (1) & (2) are both positive clearances,

CLASS OF FIT = CLEARANCE FIT.

Figure 18–1 shows dimensioned pictorials of the other two classes of fit and needed dimensions to determine (A) an interference fit and (B) a transition fit.

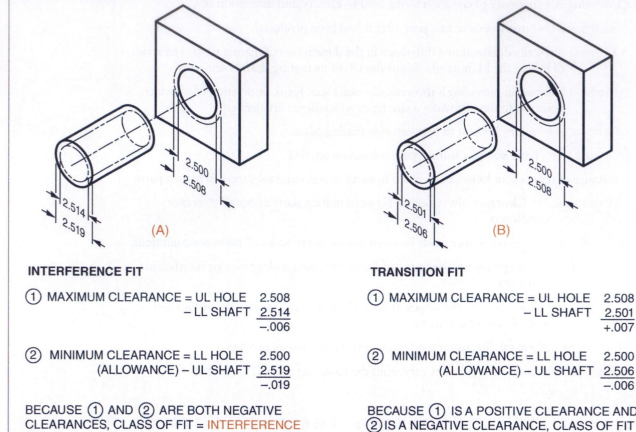

INTERFERENCE FIT

① MAXIMUM CLEARANCE = UL HOLE   2.508
                  – LL SHAFT   <u>2.514</u>
                                     –.006

② MINIMUM CLEARANCE = LL HOLE   2.500
   (ALLOWANCE) – UL SHAFT   <u>2.519</u>
                                       –.019

BECAUSE ① AND ② ARE BOTH NEGATIVE
CLEARANCES, CLASS OF FIT = INTERFERENCE
FIT

TRANSITION FIT

① MAXIMUM CLEARANCE = UL HOLE   2.508
                  – LL SHAFT   <u>2.501</u>
                                       +.007

② MINIMUM CLEARANCE = LL HOLE   2.500
   (ALLOWANCE) – UL SHAFT   <u>2.506</u>
                                       –.006

BECAUSE ① IS A POSITIVE CLEARANCE AND
② IS A NEGATIVE CLEARANCE, CLASS OF FIT =
TRANSITION FIT

FIGURE 18–1
Interference and transition fits.

ASSIGNMENT A–UNIT 18

STUDENT'S NAME _____

BASIC HOLE SIZE	BASIC SHAFT SIZE	TOLERANCE HOLE	TOLERANCE SHAFT	LIMITS HOLE UPPER/LOWER	LIMITS SHAFT UPPER/LOWER	CLEARANCE	ALLOWANCE	CLASS OF FIT*	PROBLEM NUMBER
.628	.625		+.001 / −.001	.630 / .628	.626 / .624			CL	1
1.000	.997	+.001 / −.001	+.001 / −.001	1.001 / .999	—		+.001		2
1.000		+.002 / −.003	+.001 / −.001	___ / .997	1.004 / 1.002	+.001			3
2.235	2.240	+.002 / −.002	+.002 / −.002	—	—		−.000		4
	4.0012	+.0002 / −.0002	+.0002 / −.0002	4.0008 / 4.0004	—	−.0002			5

*CLASS OF FIT CODES:
CLEARANCE = CL
INTERFERENCE = INTER
TRANSITION = TRANS

ASSIGNMENT 18A ALLOWANCE & CLASS OF FIT CHART
DIRECTIONS: FILL IN ALL EMPTY SPACES IN THE ABOVE CHART.
USE GIVEN DIMENSIONS TO DETERMINE MISSING VALUES.

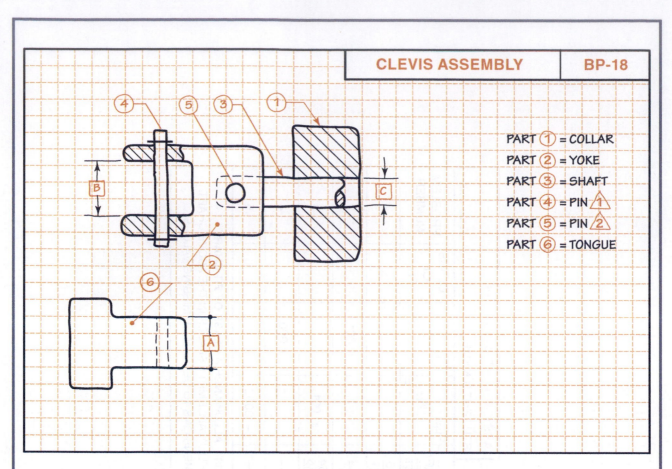

CLEVIS ASSEMBLY | **BP-18**

PART ① = COLLAR
PART ② = YOKE
PART ③ = SHAFT
PART ④ = PIN ⚠1
PART ⑤ = PIN ⚠2
PART ⑥ = TONGUE

ASSIGNMENT B—UNIT 18: CLEVIS ASSEMBLY (BP-18)

Student's Name _____

DIRECTIONS: Refer to the sketched clevis assembly blueprint above when answering the questions below.

1. If the Shaft Ø, Part ③, has an UL of 6.202, what would the limits of Collar Ø [C] need to be to have

   an allowance of −.004 with a maximum clearance of −.006? What would this class of fit be called?

   1._____

   _____

2. If Pin ⚠1 is to have an allowance of +.003, what would its UL be to fit into the 4.002 LL Yoke Holes?

   2._____

3. If Part ⑤ needs to be press-fitted into Part ② and through Part ③, what would be the UL and LL size to create an interference of +.004 when the Hole LL is 3.998?

   3._____

4. If the Yoke Mouth Opening [B] has a UL size of 10.506 with tolerance of ±.004, what would the UL

   of Tongue Dimension [A] have to be to create an allowance of +.003?

   4. _____

Representing and Dimensioning External Screw Threads

Screw threads are used widely (1) to fasten two or more parts securely in position, (2) to transmit power, such as a feed screw, on a machine, (3) to move a scale on an instrument used for precision measurements, and (4) to adjust the alignment of parts.

PROFILE OF UNIFIED AND THE AMERICAN NATIONAL THREADS

The shape or profile of the thread is referred to as the thread form. One of the most common thread forms resembles a V. Threads with an included angle of 60° originally were called Sharp V. Later, these threads were called U.S. Standard, American National, and Unified threads. Changes and improvements in the thread forms have been made over the years through standards that have been established by professional organizations. These include the National Screw Thread Commission, the American Standards Association, the U.S. Standards Institute and (currently) the American Society of Mechanical Engineers (ASME). The Unified UN thread form and sizes are designed to correct production problems related to tolerances and classes of fit. The Unified Thread Series was agreed upon by Canada, the United States, and the United Kingdom.

The most commonly used thread forms are the Unified and the American National. UN threads are mechanically interchangeable with the former American National threads of the same size and pitch. The symbol UN is found on drawings to designate the Unified form, and N designates the American National form. The only difference between these two forms is in the shape of the top (crest) and the bottom (root) of the thread. Both the crest and the root of the American National thread form are flat. In contrast, the crest and root of the Unified thread form may be either flat or rounded. The root is always rounded on UNR external threads. The characteristics of the basic profile for American National and Unified screw threads are shown in Figure 19–1.

AMERICAN NATIONAL STANDARD AND UNIFIED SCREW THREADS

There are five thread forms that use the 60° included angle. These forms are represented in Figure 19–2 as (A) the Sharp V, (B) Unified, UN, (C) American Standard, N, (D) 60° Stub, and (E) ASME Straight, NPS, and Taper, NPT, pipe threads. While flat crests and roots are shown for the pipe threads, some rounding off occurs in manufacturing.

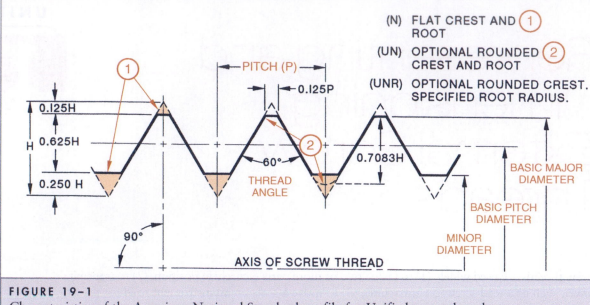

FIGURE 19-1
Characteristics of the American National Standard profile for Unified screw threads
(N, UN, and UNR).

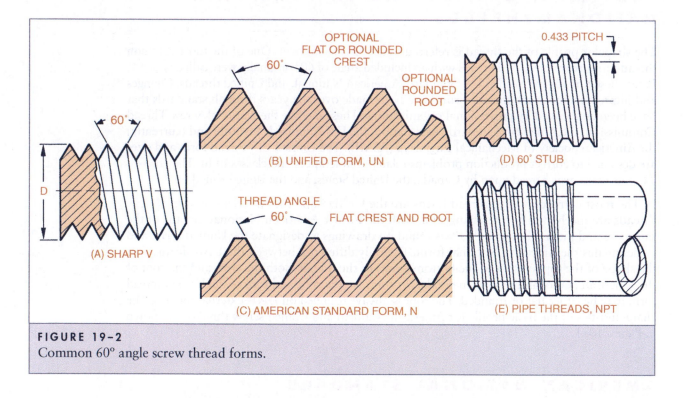

FIGURE 19-2
Common 60° angle screw thread forms.

OTHER COMMON THREAD FORMS

A square form of thread, or a modified form, is used when great power or force is needed. The square form shown in Figure 19–3A is called a **Square thread.** The more popular modified thread form has an included angle of 29° and is known as the **Acme thread,** Figure 19–3B. The **Buttress thread,** Figure 19–3C, is used where power is to be transmitted in one direction. The load-resisting flank may be inclined from 1° to 5°.

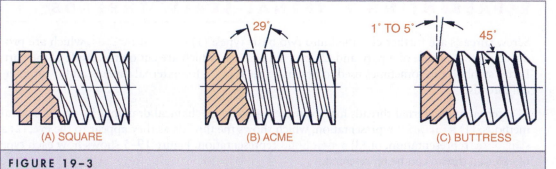

FIGURE 19–3
Square and modified thread forms.

THREAD SERIES

Each thread form has a standard number of threads per inch for a given diameter. The thread series designates the *fineness* or *coarseness* of the threads. For example, there are five basic series of 60° thread form screw threads, as follows:

1. The **coarse thread series** is designated as **UNC** for Unified Coarse; **UNR** for Unified Coarse, having a special rounded root but optional rounded crest; and **NC** for the American National coarse series threads.
2. The **fine thread series** is similarly identified by the symbols **UNF, UNRF** for external threads, and **NF.**
3. **Extra fine series threads** are designated by the symbols **UNEF, UNREF** (for external threads only), and **NEF.**
4. **Nonstandard** or **special threads** are designated by the symbol **UNS** for Unified Special and **NS** for American National standard special series threads.
5. A **constant pitch series** provides a fixed number of threads (pitch) for a range of diameters. The **8, 12,** and **16** constant pitch series are preferred. Threads in each series are designated as **UN, UNR** (for external threads), and **N.** These letters (depending on the thread form) are preceded by the number of threads in the constant pitch series.

Figure 19–4 provides a summary of the five thread series and the designations as they appear on drawings.

	Designations on Drawings		
Thread Series	**Unified System**		**American National System**
Coarse	/ 1 UNC	/ 2 UNRC	NC
Fine	UNF	UNRF	NF
Extra Fine	UNEF	UNREF	NEF
Special	UNS	UNRS	NS
Constant Pitch	/ 3 * UN	* UNR	* N

/ 1   Optional rounded crest and/or root
/ 2   Specified root radius; optional rounded crest
/ 3   * Pitch series number precedes the designation

FIGURE 19–4
Summary of thread series designations.

REPRESENTING EXTERNAL SCREW THREADS

Screw threads are further classified into two basic types: (1) **external threads,** which are produced on the outside of a part, and (2) **internal threads,** which are cut on the inside of the part. Letter symbols are sometimes used to designate the type: **A** for external threads and **B** for internal threads.

Internal and external threads may be represented on mechanical drawings by one of several methods: (1) a **pictorial** representation, which shows the threads as they appear to the eye, (2) a **schematic** representation, or (3) a **simplified** representation. Figure 19–5 shows how each type of external thread can be represented.

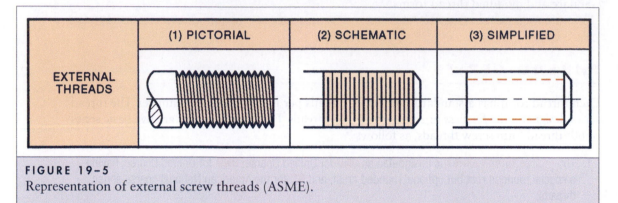

FIGURE 19-5
Representation of external screw threads (ASME).

DIMENSIONING EXTERNAL SCREW THREADS

The representation of each thread is accompanied by a series of dimensions, letters, and numbers, which, when combined, give full specifications for cutting and measuring the threads. The standard practices recommended by the ASME for specifying and dimensioning external screw threads are shown in Figure 19–6.

While the specifications as noted in Figure 19–6 give all the information needed to describe a screw thread, the seven items are not always used in the notation on a drawing. For example, the

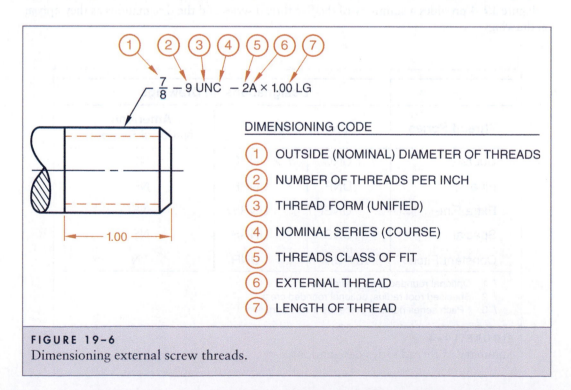

FIGURE 19-6
Dimensioning external screw threads.

thread class 5 may be covered by a general note that applies to all threaded parts. Such a note may appear elsewhere on the drawing. In other cases, the length of thread 7 sometimes appears as a dimension instead of appearing in the thread notation.

PIPE THREAD REPRESENTATION

In addition to the screw threads just covered, there are three other common types of ASME threads that are used on pipes. One type is the straight pipe thread. This type of thread is noted on a drawing by the diameter of the thread followed by NPS. Thus, the notation 1 - NPS, indicates that (a) the part contains the standard number of threads in the pipe series for a 1″ diameter and (b) the threads are straight.

In the second type of pipe thread, the threads are cut along a standard taper. The tapered pipe threads are used where a tight, leakproof seal is needed between the parts that are joined. Such threads are called National Pipe Taper threads, designated by NPT. They are specified on pictorial, schematic, or simplified drawings by the diameter and the type. The notation, 1 - NPT, means (a) that there are the standard number of threads on the part for the 1″ diameter in the pipe thread series and (b) both the internal and external threads are cut on the standard pipe taper.

A third type of pipe thread is used on drawings of parts for extremely high pressure applications. Dryseal pressure-tight threads are identified by the addition of the letter F after the T (taper) or S (straight) thread form designations. For example, NPTF designates threads in the Dryseal taper series; NPSF designates threads in the Dryseal straight pipe thread series. See Figure 19-7.

While a number of different forms and thread series have been described, only the commonly used ASME, Unified System, and International Standards Organization (ISO) Metric Thread System are applied in this basic text.

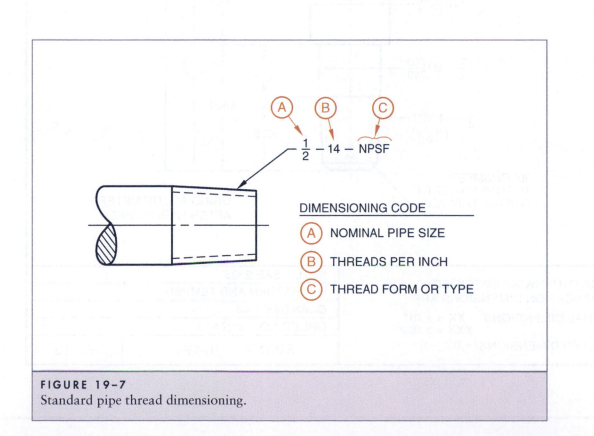

FIGURE 19-7
Standard pipe thread dimensioning.

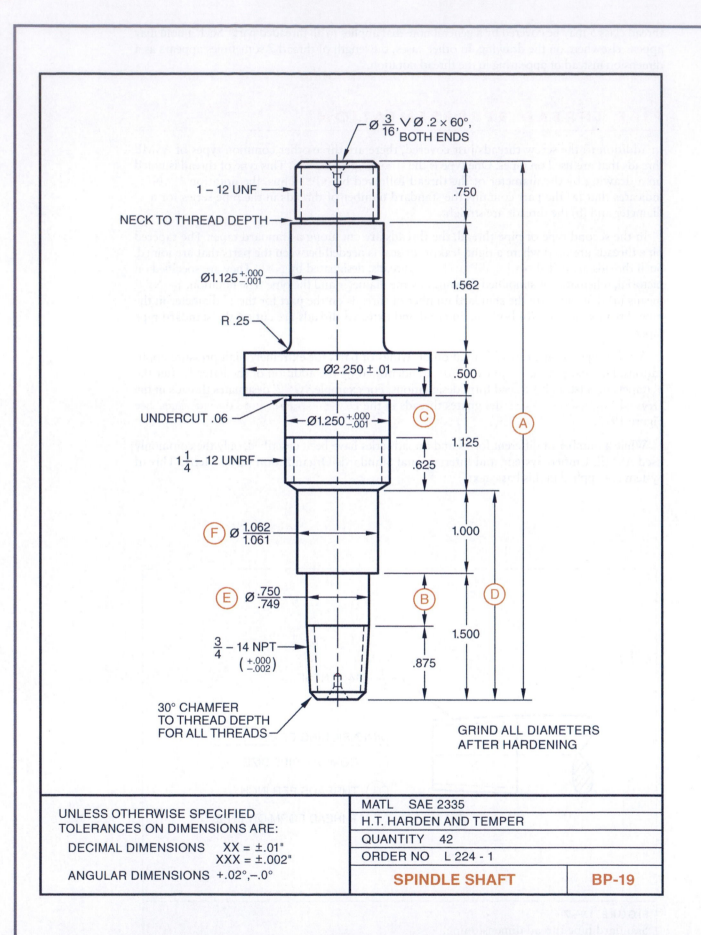

Ø $\frac{3}{16}$, ⌵ Ø .2 × 60°, BOTH ENDS

1 – 12 UNF

NECK TO THREAD DEPTH

Ø1.125 $^{+.000}_{-.001}$

R .25

Ø2.250 ± .01

UNDERCUT .06

Ø1.250 $^{+.000}_{-.001}$

C

1$\frac{1}{4}$ – 12 UNRF

F Ø $\frac{1.062}{1.061}$

E Ø $\frac{.750}{.749}$

$\frac{3}{4}$ – 14 NPT ($^{+.000}_{-.002}$)

30° CHAMFER TO THREAD DEPTH FOR ALL THREADS

.750

1.562

.500

A

1.125

.625

1.000

B   D

1.500

.875

GRIND ALL DIAMETERS AFTER HARDENING

UNLESS OTHERWISE SPECIFIED TOLERANCES ON DIMENSIONS ARE:	MATL SAE 2335
DECIMAL DIMENSIONS XX = ±.01" XXX = ±.002"	H.T. HARDEN AND TEMPER
ANGULAR DIMENSIONS +.02°,–.0°	QUANTITY 42
	ORDER NO L 224 - 1
	SPINDLE SHAFT BP-19

ASSIGNMENT—UNIT 19: SPINDLE SHAFT (BP-19)

Student's Name _____

1. What material is used for the part?
1. _____

2. Give the basic overall length Ⓐ of the shaft.
2. _____

3. What system of representation is used for the threaded portions?
3. _____

4. At how many places are threads cut?
4. _____

5. Start at the bottom of the part and give all the thread diameters.
5. _____
_____
_____

6. Name the three thread series that the letters **UNC, UNRF,** and **NPT** specify.
6. UNC _____
UNRF _____
NPT _____

7. How many threads per inch are to be cut on the 3/4", 1 1/4", and 1" diameters?
7. 3/4" _____
1 1/4" _____

8. Give basic dimensions, Ⓑ, Ⓒ, and Ⓓ.
8. Ⓑ _____
Ⓒ _____
Ⓓ _____

9. Give the upper- and lower-limit diameter for the 3/4" threaded portion.
9. Upper _____
Lower _____

10. What is the length of the 3/4" threads? The 1 1/4" threads?
10. 3/4" _____
1 1/4" _____

11. Give the upper- and lower-limit dimensions of the Ⓕ diameter portion.
11. Upper _____
Lower _____

12. At what angle are the chamfers at the starting end of each thread cut?
12. _____

13. What tolerance is specified for angular dimensions?
13. _____

14. What is the upper- and lower-limit of size on the 1 1/8" diameter?
14. Upper _____
Lower _____

15. How long is the part of the shaft that has the 1 1/4" – 12 thread?
15. _____

16. Give the angle of the countersink.
16. _____

17. What is the largest diameter that the 2.25" diameter can be machined to?
17. _____

18. Give the diameter to which the center holes are countersunk.
18. _____

19. How deep is the undercut on the 1 1/4" diameter?
19. _____

20. Name the final machining operation for all diameters after the part is heat-treated.
20. _____

21. Classify the tolerances for ground diameters Ⓔ and Ⓕ.
21. Ⓔ _____
Ⓕ _____

20 Representing and Specifying Internal and Left-Hand Threads

REPRESENTATION OF INTERNAL THREADS

One of the most practical and widely used methods of producing internal threads is to cut them with a round, formed, thread-cutting tool called a tap. The tap is inserted in a hole that is the same diameter as the root diameter of the thread to be produced. As the tap with its multiple cutting edges is turned, it advances into the hole to cut a thread. The thread is a specified size and shape to correspond with a mating threaded part. This process is called tapping, and the hole that is threaded is called a tapped hole. When the hole is a drilled hole, the drill is called a tap drill.

A part may be threaded internally throughout its entire length or to a specified depth, either by tapping or another machining process. Some threads are bottomed at the same depth as the tap drill. In other cases, the tap drill hole may be deeper. Pictorial, schematic, and simplified forms of representing such threads are used in Figure 20–1. Section (A) illustrates through threads; Section (B) illustrates threads in a blind hole.

DIMENSIONING A THREADED HOLE

The same system of dimensioning that is used to specify external threads applies to internal threads. However, in the case of internal threads, the thread length is specified as the depth of thread. The recommended dimensioning of a tapped hole is illustrated in Figure 20–2.

THREAD CLASS: UNIFIED AND AMERICAN NATIONAL FORM THREADS

The accuracy between an external and an internal thread is specified in trade handbooks and manufacturer's tables as a thread class. In the old classes of thread fits, there were equal allowances (tolerances) on the pitch diameters of both the external and internal threads. Today, the new standards require a larger pitch diameter tolerance on the internal thread.

There are three regular classes of fits for external and internal threads. Each of the three classes is distinguished from one another by the amount of clearance between the sides, crests, and roots of the internal and external threads.

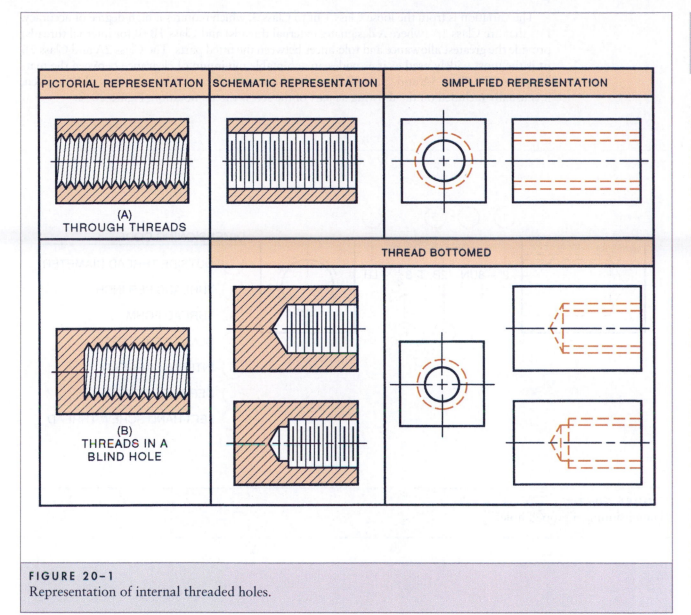

PICTORIAL REPRESENTATION	SCHEMATIC REPRESENTATION	SIMPLIFIED REPRESENTATION
(A) THROUGH THREADS		
(B) THREADS IN A BLIND HOLE	THREAD BOTTOMED	

FIGURE 20–1
Representation of internal threaded holes.

The variation is from the loose Class 1 fit to Class 3, which requires a high degree of accuracy. **Fits** that are Class 1A (where A designates external threads) and Class 1B (B for internal threads) provide the greatest allowance and tolerances between the fitted parts. The Class 2A and Class 2B fit is the most widely used as it provides an acceptable minimum of clearance between the mating parts. The Class 3A and Class 3B fit is used for great accuracy. When the class of fit is given, it is usually included on the drawing as part of the screw thread notation, Figure 20–2.

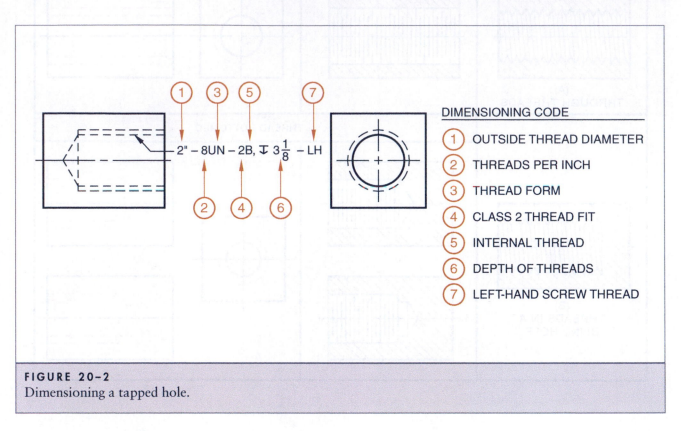

DIMENSIONING CODE

(1) OUTSIDE THREAD DIAMETER

(2) THREADS PER INCH

(3) THREAD FORM

(4) CLASS 2 THREAD FIT

(5) INTERNAL THREAD

(6) DEPTH OF THREADS

(7) LEFT-HAND SCREW THREAD

FIGURE 20–2
Dimensioning a tapped hole.

SPECIFYING LEFT-HAND THREADS

Screw threads are cut either right-hand or left-hand depending on the application, Figure 20–3. As the terms imply, a right-hand thread is advanced by turning clockwise ⌒ or to the right; the left-hand thread is advanced by turning counterclockwise ⌒ or to the left.

No notation is made on a drawing for right-hand threads, as this direction of thread is assumed unless otherwise noted. However, the letters **LH** are included on the drawing specifications to indicate when a left-hand thread is required, as shown in Figure 20–2 by the **LH** location at ⑦.

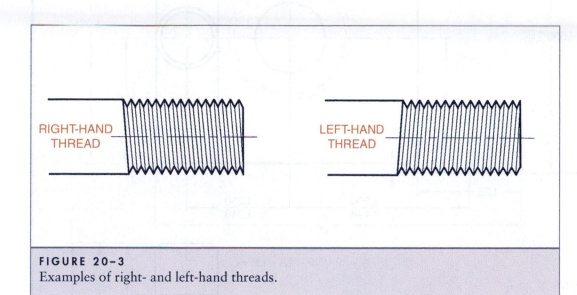

FIGURE 20-3
Examples of right- and left-hand threads.

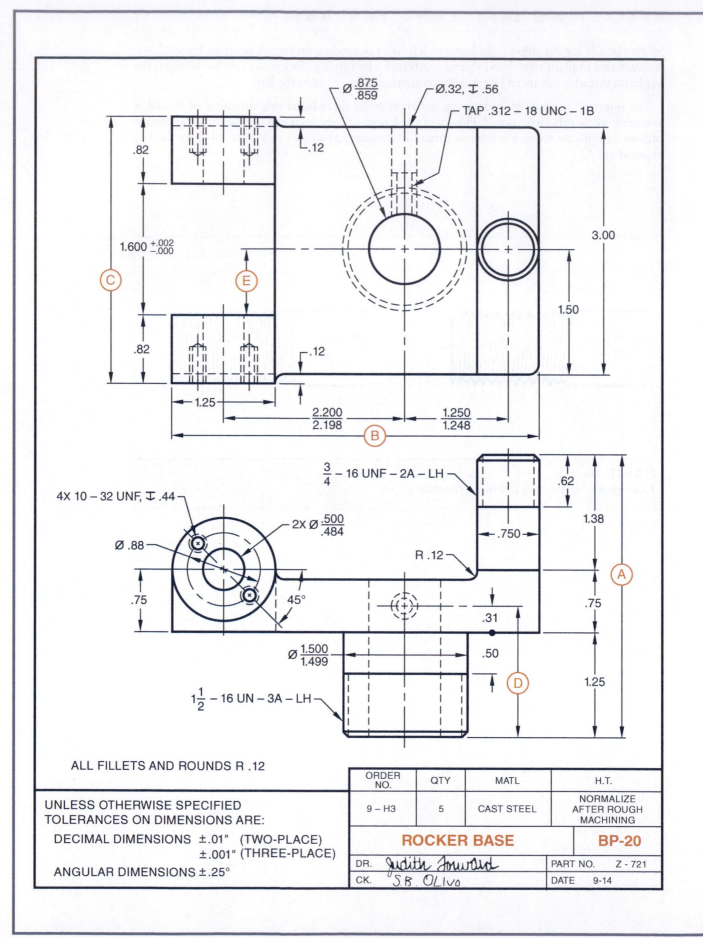

Ø .875 / .859

Ø .32, ⊔ .56

TAP .312 – 18 UNC – 1B

.82

.12

1.600 +.002 / -.000

3.00

1.50

.82

.12

1.25

2.200 / 2.198

1.250 / 1.248

C

E

B

3/4 – 16 UNF – 2A – LH

4X 10 – 32 UNF, ⊔ .44

2X Ø .500 / .484

Ø .88

R .12

.62

1.38

.750

.75

45°

A

.75

.31

Ø 1.500 / 1.499

.50

D

1.25

1½ – 16 UN – 3A – LH

ALL FILLETS AND ROUNDS R .12

ORDER NO.	QTY	MATL	H.T.
9 – H3	5	CAST STEEL	NORMALIZE AFTER ROUGH MACHINING

UNLESS OTHERWISE SPECIFIED
TOLERANCES ON DIMENSIONS ARE:

DECIMAL DIMENSIONS ±.01" (TWO-PLACE)
±.001" (THREE-PLACE)

ANGULAR DIMENSIONS ±.25°

ROCKER BASE

BP-20

DR.	Judith Forward	PART NO.	Z - 721
CK.	S.B. Olivo	DATE	9-14

ASSIGNMENT—UNIT 20: ROCKER BASE (BP-20)

Student's Name _____

1. What material is specified?

2. What tolerances are allowed on:

   (a) Decimal dimensions?

   (b) Fractional dimensions?

   (c) Angular dimensions?

3. What heat-treating process is required after rough machining?

4. Determine overall basic dimensions Ⓐ Ⓑ Ⓒ.

5. What is the radius of all fillets and rounds?

6. Compute dimensions Ⓓ and Ⓔ.

7. How many holes are .500?

8. How many external threads are to be cut?

9. What thread series is used for the internal and external threads?

10. What does 3/4 - 16UNF - 2A - LH mean?

11. What are the upper- and lower-limits of the .750″ diameter portion?

12. What does 1 1/2 - 16UN - 3A - LH mean?

13. What does .312 - 18UNC - 1B mean?

14. What is the lower-limit diameter of the unthreaded 1 1/2″ diameter portion?

15. Determine the length of the 1 1/2″ - 16 threaded portion.

16. How many holes are to be threaded 10 - 32UNF?

17. How deep are the holes to be threaded?

18. Give the angle to the horizontal at which the 10 - 32UNF holes are to be drilled.

19. Give the diameter of the circle on which the 10 - 32UNF tapped holes are located.

20. Identify the system of dimensioning used on the drawing of the **rocker base**.

21. Change the decimal and angular tolerances so they apply in one direction (+) only. Then, determine the upper- and lower-limit dimensions for Ⓐ, Ⓑ, Ⓒ, Ⓓ, Ⓔ, and the 45° angle.

1. _____
2. (a) _____
   (b) _____
   (c) _____
3. _____
4. Ⓐ _____
   Ⓑ _____
   Ⓒ _____
5. _____
6. Ⓓ _____
   Ⓔ _____
7. _____
8. _____
9. _____
10. 3/4 _____
    16UNF _____
    2A _____
    LH _____
11. Upper _____
    Lower _____
12. 1 1/2 _____
    16UN _____
    3A _____
    LH _____
13. .312 _____
    18UNC _____
    1B _____
14. _____
15. _____
16. _____
17. _____
18. _____
19. _____
20. _____
21. Ⓐ _____
    Ⓑ _____
    Ⓒ _____
    Ⓓ _____
    Ⓔ _____
    45° _____

Dimensioning Tapers and Machined Surfaces

Blueprints of parts that uniformly change in size along their length show that the part is tapered. On a round piece of work, the **taper** is the difference between the diameter at one point and a diameter farther along the length. The drawing may give the large and small diameters at the beginning and end of the taper or one diameter and the length of the taper, Figure 21–1. The taper is usually specified on drawings where dimensions are given in inches by a note that gives the **TAPER PER FOOT** or the **TAPER PER INCH**. For example, a taper of one-half inch per foot is expressed:

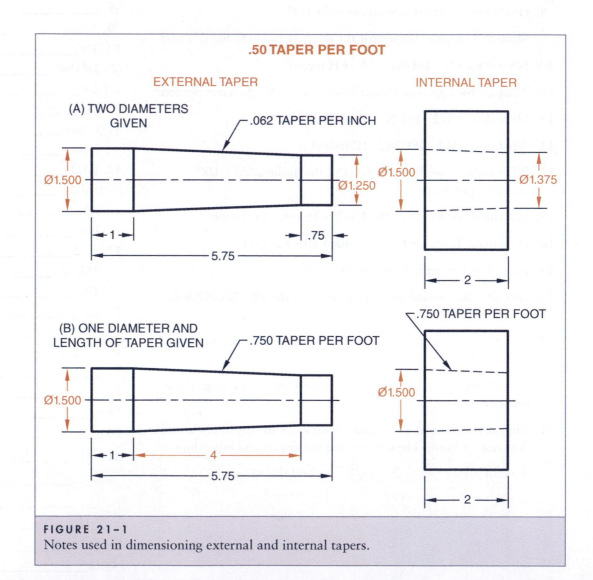

.50 TAPER PER FOOT

EXTERNAL TAPER   INTERNAL TAPER

(A) TWO DIAMETERS GIVEN   .062 TAPER PER INCH

Ø1.500   Ø1.250   Ø1.500   Ø1.375

1   5.75   .75

(B) ONE DIAMETER AND LENGTH OF TAPER GIVEN   .750 TAPER PER FOOT   .750 TAPER PER FOOT

Ø1.500   Ø1.500

1   4   5.75   2   2

FIGURE 21–1
Notes used in dimensioning external and internal tapers.

The newest ASME standard for dimensioning conical (cone-shaped) tapered surfaces recommends the use of a ratio followed by the taper symbol (- ▷). The ratio represents the difference in diameter in relation to the length of the taper. For example, a part that is dimensioned in customary units (CIU) of .25″ in 5.00″ has a CIU taper ratio per unit (inch) length of .25:5.00 and is shown on a drawing as - ▷ 0.2:1. Similarly, a conical part with a metric diameter difference of 20 mm and length of 100 mm has a taper ratio (TR) of 20.0:100.0 or 1:5 - ▷ (metric).

CALCULATING OUTSIDE AND INSIDE TAPER MEASUREMENTS

To find the total taper (T_t) when two diameters of a taper are given, subtract the diameter of the small end (d) from the diameter of the large end (D) of the taper. $T_t = D - d$. Applied to the external taper in Figure 21–1(A), $T_t = 1.500″ - 1.250″ = .250″$.

To find the total taper (T_t) when one diameter, the taper per foot (T_{pf}), and the length of taper (ℓ) are given, follow two steps.

STEP 1 Divide the taper per foot by twelve to get the taper per inch.

STEP 2 Multiply the taper per inch by the length of the taper in inches.

In the case of the internal taper, Figure 21–1(B), the taper per inch (T_{pi}) = .750 ÷ 12 = .0625″. The total taper (T_t) = .0625″ × 2″ = .125″.

FINISHED SURFACES

Surfaces that are machined to a smooth finish and to accurate dimensions are called finished surfaces. These surfaces were identified by the former 60° V ANSI symbol, the √ symbol that was accepted by the Canadian Standards Association (CSA) and the ASME. Presently, the symbol √ is used to indicate that material is to be removed by machining the designated surface.

The bottom of the finish symbol touches the surface to be machined. Figure 21–2 shows a steel forging. This same forging as it looks when machined is illustrated in Figure 21–3. The working drawing, Figure 21–4, shows how the finish marks are placed. Note that these marks may appear on the hidden edge lines as well as on the object lines, with the exception of the small drilled holes. On large holes that require accurate machining, the √ is used once again. Extension lines and/or leaders are also used with the finish symbol to identify surfaces that are to be machined.

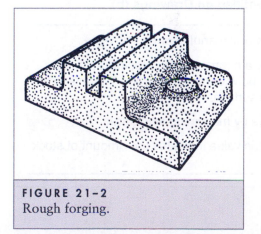

FIGURE 21–2
Rough forging.

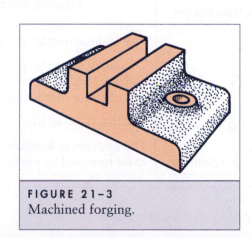

FIGURE 21–3
Machined forging.

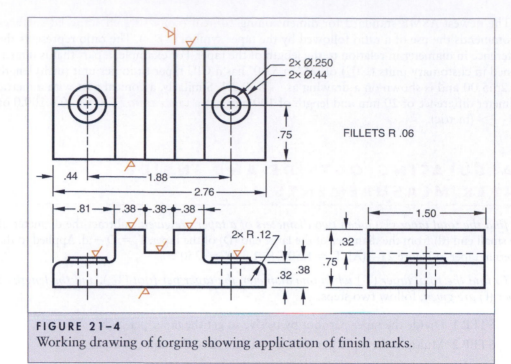

FIGURE 21-4
Working drawing of forging showing application of finish marks.

FINISH ALL OVER

When a casting, forging, or welded part is to be finished all over, the drawing is simplified by omitting the finish symbols and adding the note: **FINISH ALL OVER.** The abbreviation **FAO** may be used in place of the phrase.

ASME SYMBOLS FOR SURFACE FINISH PROCESSES

Four basic ASME symbols for denoting the process by which a surface may be finished are shown in Figure 21–5, column (a). A brief description of the process that each symbol identifies on a drawing is given in column (b).

ASME Surface Finish Symbol (a)	General Process Identified on Drawings (b)
✓	The surface may be produced by any manufacturing process.
⊄	The surface is to be produced by a nonmachining process such as casting, forging, welding, rolling, powder metallurgy, etc. There is to be no subsequent machining of the surface.
✓	Material removal from the surface by machining is required.
.05" ✓ or 1.5mm ✓	The addition of a decimal or metric value indicates the amount of stock to be removed by machining.

FIGURE 21-5
Basic ASME surface finish symbols.

SURFACE TEXTURE MEASUREMENT TERMS AND SYMBOLS

Surface texture, surface finish, surface roughness, and surface characteristics are generally used interchangeably in the shop and laboratory. However, application of these terms to drawings must follow precise ASME standards. Surface texture symbols and values provide specific standards against which finished parts may be accurately produced, uniformly inspected, and measured. Common surface texture characteristics are identified in Figure 21–6, and a description of basic terms follows.

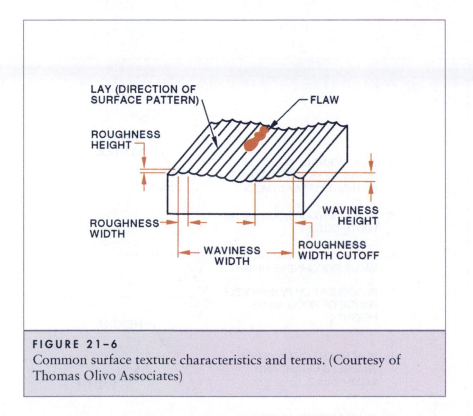

FIGURE 21-6
Common surface texture characteristics and terms. (Courtesy of Thomas Olivo Associates)

Surface texture. As applied to drawings, surface texture denotes repetitive or random deviations from a nominal (specified) surface. These variations form the **pattern** of the surface.

Surface finish. The symbol √ and a surface roughness measurement value are used to indicate the roughness height in microinches (μin) or micrometers (μm). For instance, if the roughness height limit for a machined surface is 125 microinches (125 μin), this measurement is shown on a drawing as ¹²⁵√ .

Microinch or micrometer value. The microinch (μin) and micrometer (μm) are the base units of surface roughness measurement. The prefix micro indicates a millionth part of an inch or meter, respectively.

Lay. Every machining and superfinishing process produces lines that form six basic patterns. The patterns are identified on drawings using **lay symbols.** Symbols that give the lay direction are: (⊥) perpendicular, (∥) parallel, (X) angular, (M) multidirectional, (C) circular, and (R) radial.

Waviness. A series of waves in a surface may be produced by machining, stresses or defects in a material, heat treating, handling, or other processing. Waviness is measured by height (in thousandths of an inch) and width (in other decimal parts of an inch). **Waviness width** refers to the measurement between successive peaks or valleys.

Flaws. Defects in a surface such as scratches, burrs, casting, forging, machining, or other imperfections are referred to as flaws.

REPRESENTATION AND INTERPRETATION OF SURFACE TEXTURE MEASUREMENTS

The ASME surface texture system requires a series of measurement values to be used with the symbol. The values specify the degree of accuracy to which surface irregularities (roughness) are to be measured. The pattern of marks produced by machining (lay), the spacing of the irregularities (waviness), and scratches and other surface imperfections (flaws) are represented by still other symbols and measurement values. The placement of the surface texture roughness and waviness symbols and values is shown graphically in Figure 21–7.

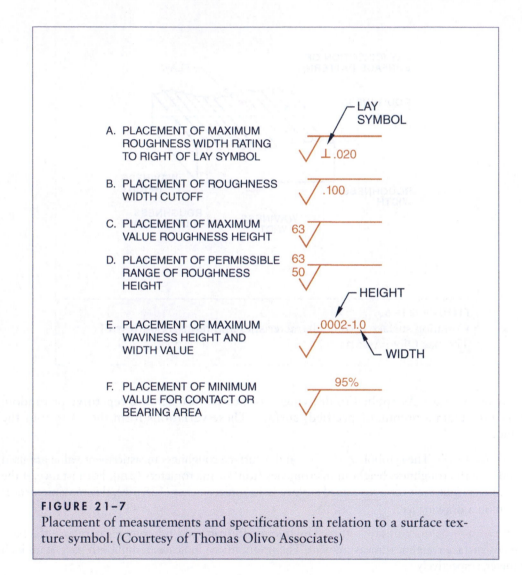

A. PLACEMENT OF MAXIMUM ROUGHNESS WIDTH RATING TO RIGHT OF LAY SYMBOL

LAY SYMBOL

⊥ .020

B. PLACEMENT OF ROUGHNESS WIDTH CUTOFF

.100

C. PLACEMENT OF MAXIMUM VALUE ROUGHNESS HEIGHT

63

D. PLACEMENT OF PERMISSIBLE RANGE OF ROUGHNESS HEIGHT

63
50

E. PLACEMENT OF MAXIMUM WAVINESS HEIGHT AND WIDTH VALUE

HEIGHT

.0002-1.0

WIDTH

F. PLACEMENT OF MINIMUM VALUE FOR CONTACT OR BEARING AREA

95%

FIGURE 21–7
Placement of measurements and specifications in relation to a surface texture symbol. (Courtesy of Thomas Olivo Associates)

Figure 21–8(A) illustrates the way lay symbols and surface texture specifications should appear on a drawing. Interpretation of the lay direction and each surface texture measurement is given at (B).

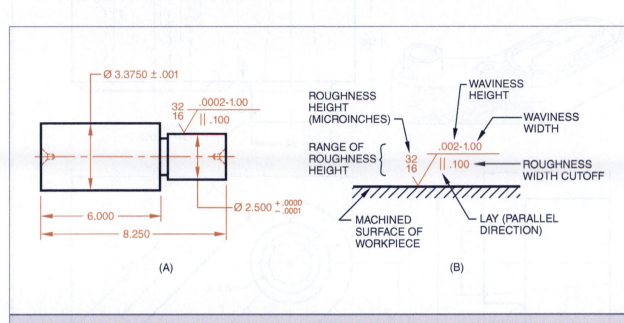

(A)   (B)

FIGURE 21–8
Representation and interpretation of surface texture symbols and measurement values. (Courtesy of Thomas Olivo Associates)

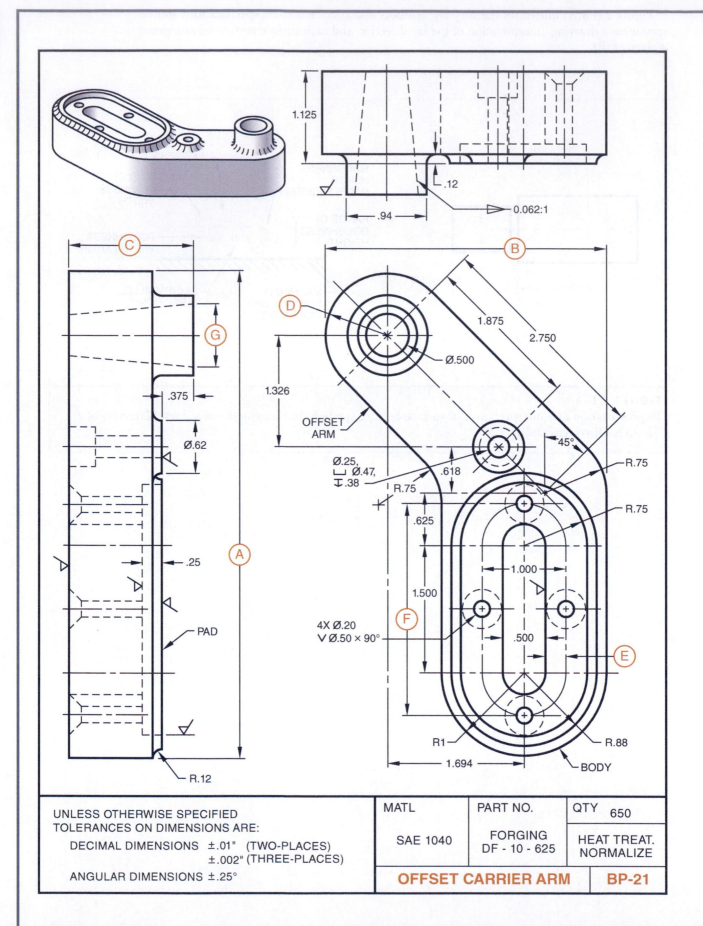

1.125

.12

.94

0.062:1

C

G

.375

Ø.62

.25

A

PAD

R.12

B

D

1.875

2.750

Ø.500

1.326

OFFSET
ARM

45°

Ø.25,
Ø.47,
.38

.618

R.75

R.75

R.75

.625

R.75

1.000

1.500

4X Ø.20
Ø.50 × 90°

F

.500

E

R1

R.88

1.694

BODY

MATL	PART NO.	QTY 650
SAE 1040	FORGING DF - 10 - 625	HEAT TREAT. NORMALIZE

UNLESS OTHERWISE SPECIFIED
TOLERANCES ON DIMENSIONS ARE:

DECIMAL DIMENSIONS ±.01" (TWO-PLACES)
±.002" (THREE-PLACES)

ANGULAR DIMENSIONS ±.25°

OFFSET CARRIER ARM   **BP-21**

ASSIGNMENT—UNIT 21: OFFSET CARRIER ARM (BP-21)

Student's Name _____

1. Name the three views.

2. What is the part number?

3. How many surfaces are to be machined?

4. Give the angle at which the arm is offset from the body.

5. What is the basic center-to-center distance between the holes in the offset arm?

6. Find the basic length of the elongated slot.

7. What is the basic overall length of the pad?

8. How is the counterbored hole specified?

9. What diameter drill is used for the countersunk holes?

10. Give the angle of the countersunk holes.

11. What tolerance is allowed on:

    (a) Decimal dimensions?

    (b) Angular dimensions?

12. Find the maximum overall thickness Ⓒ.

13. Compute the basic diameter at the large end of the tapered hole Ⓖ.

14. What is dimension Ⓓ?

15. What is the basic distance Ⓔ?

16. Give basic center-to-center distance Ⓕ.

17. What is the basic overall height Ⓐ?

18. Find the basic overall width Ⓑ.

19. If the elongated slot is machined .48, would it be over, under, or within the specified limits?

20. What heat treatment is required?

21. What classification of tolerances applies to all dimensions?

22. Indicate the meaning of the ⤾ symbol on BP-21.

23. Give (a) the dimensions and (b) state surface texture characteristics for, Ⓘ, Ⓙ, Ⓚ, and Ⓛ.

Ⓘ—— 64
          Ⓙ   Ⓛ
      32 .001 − 2.00
      √× .300 ——Ⓚ

24. What does the lay symbol × mean?

1. _____

_____

_____

2. _____

3. _____

4. _____

5. _____

6. _____

7. _____

8. _____

_____

9. _____

10. _____

11. (a) _____

(b) _____

12. Ⓒ = _____

13. Ⓖ = _____

14. Ⓓ = _____

15. Ⓔ = _____

16. Ⓕ = _____

17. Ⓐ = _____

18. Ⓑ = _____

19. _____

20. _____

21. _____

22. _____

_____

_____

23. Dimension (a)

Ⓘ = _____

Ⓙ = _____

Ⓚ = _____

Ⓛ = _____

Characteristic (b)

Ⓘ = _____

Ⓙ = _____

Ⓚ = _____

Ⓛ = _____

24. _____

22 Dimensioning with Shop Notes

The draftsperson often resorts to the use of notes on a drawing to convey to the mechanic all the information needed to make a part. Notes such as those used for drilling, reaming, counterboring, or countersinking holes are added to ordinary dimensions.

A note may consist of a very brief statement at the end of a leader or it may be a complete sentence that gives an adequate picture of machining processes and all necessary dimensions. A note is found on a drawing near the part to which it refers. This unit includes the types of machining notes that are found on drawings of knurled surfaces, chamfers, grooves, and keyways. A sample of a typical change note is also given.

DIMENSIONING KNURLED SURFACES

The term knurl refers to a raised diamond-shaped surface or straight-line impression in the surface of a part, Figure 22–1A. The dimensions and notes that furnish sufficient information for the technician to produce the knurled part are also given, Figure 22–1B and Figure 22–1C. The pitch of the knurl, which gives the number of teeth per linear inch, is the size. The standard pitches are: coarse (14P), medium (21P), and fine (33P).

Newer ASME standards for knurls and knurling use the letter P to precede the pitch. Where parts are to be press-fit, the minimum acceptable (toleranced) diameter is indicated on the drawing. A knurled surface is represented by diamond or straight-line pattern symbols or by

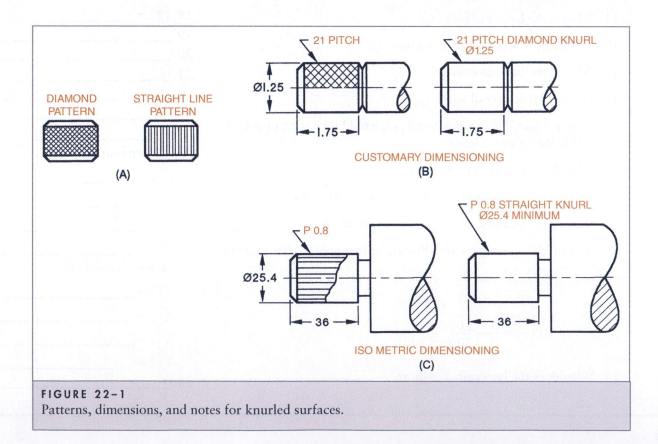

FIGURE 22–1
Patterns, dimensions, and notes for knurled surfaces.

using the designation of the knurl as straight or diamond. Common methods of representing knurls are shown in Figure 22–1B and Figure 22–1C. A partial pattern is shown where the symbol clarifies a drawing.

The pitch of a knurl is usually stated on drawings in terms of the number of teeth, as shown in Figure 22–1B, or fractional part per millimeter, Figure 22–1C. The pitch is followed by the type of knurl.

DIMENSIONING CHAMFERS AND GROOVES

When a surface must be cut away at a slight bevel or have a groove cut into it, the drawing or blueprint gives complete machining information in the form illustrated in Figure 22–2 and Figure 22–3. The angle is dimensioned as an included angle when it relates to a design requirement.

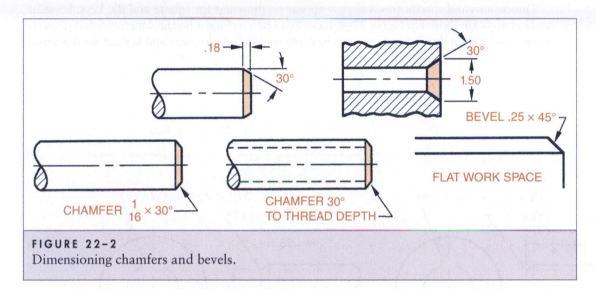

FIGURE 22–2
Dimensioning chamfers and bevels.

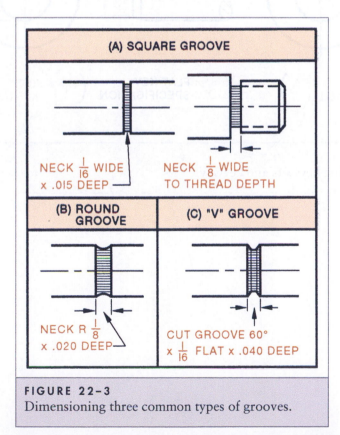

FIGURE 22–3
Dimensioning three common types of grooves.

The process of producing a groove is often referred to as **undercutting** or **necking.** The simplified ASME method of dimensioning a round or square (plain) undercut is to give the width followed by the diameter. For example, a note such as .12 × ⌀ 1.25 indicates a groove width of .12″ and a 1.25″ diameter at the bottom of the undercut. If a round groove is required, the radius is one-half the width dimension.

KEYS, KEYWAYS, AND KEYSEATS

Parts that are assembled and disassembled in a particular position to transmit motion from one member to another are **keyed** together. This means a square or flat piece of steel is partly seated in a shaft or groove called a **keyseat.** The key extends an equal distance into a corresponding groove (**keyway**) in a second member (**hub**). Another general type of key is semicircular in shape. It fits into a round machined groove in the shaft. This type is known as a **Woodruff key.**

Dimensions and specifications as they appear on drawings for square and flat keys, keyseats, and keyways are shown in Figure 22–4. Drawings often include a height dimension that gives the measurement from the circumference to the depth to which the keyway and keyseat are machined.

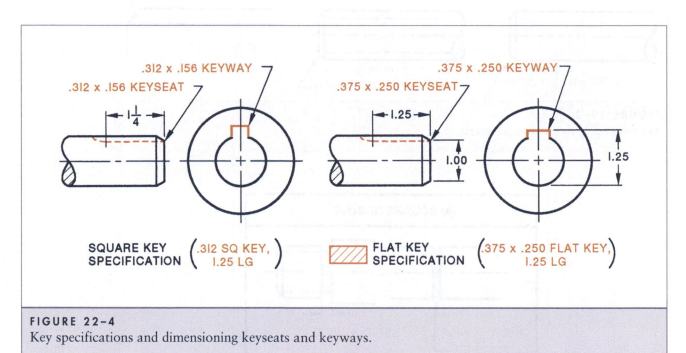

FIGURE 22–4
Key specifications and dimensioning keyseats and keyways.

CHANGE NOTES

The specifications and dimensions of parts are frequently changed on working drawings. An accurate record is usually made on the tracing and blueprint to indicate the nature of the change, the date, and who made the changes.

Changes that are minor in nature may be made without altering the original lines on the drawing. One of the easiest ways of making such changes is shown in Figure 22–5.

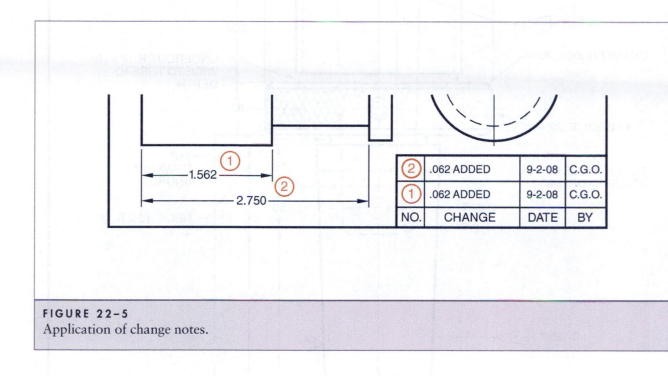

NO.	CHANGE	DATE	BY
2	.062 ADDED	9-2-08	C.G.O.
1	.062 ADDED	9-2-08	C.G.O.

FIGURE 22–5
Application of change notes.

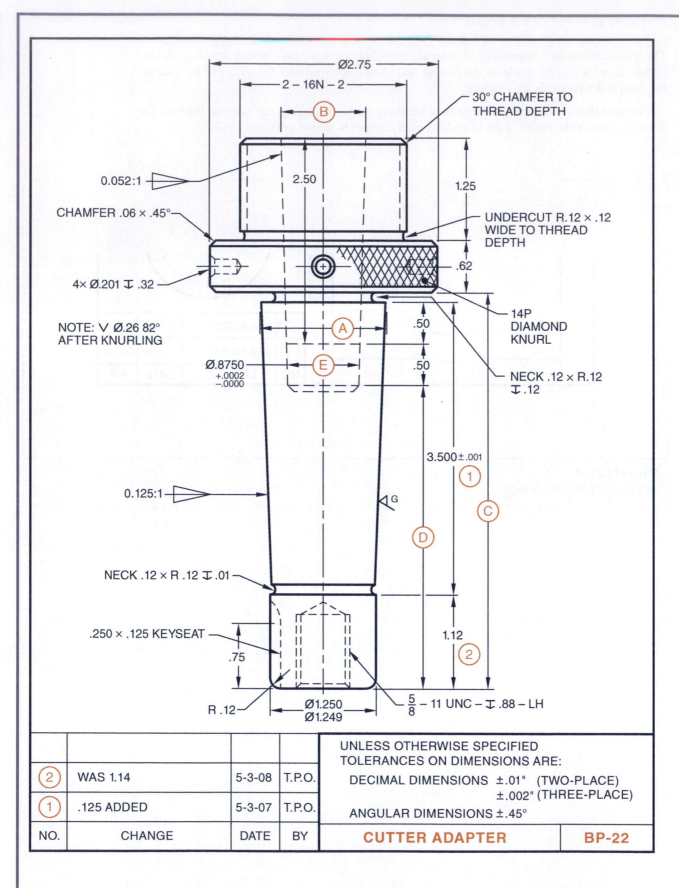

Ø2.75

2 – 16N – 2

Ⓑ

30° CHAMFER TO
THREAD DEPTH

0.052:1

2.50

1.25

CHAMFER .06 × .45°

UNDERCUT R.12 × .12
WIDE TO THREAD
DEPTH

4× Ø.201 ⊔ .32

.62

NOTE: ∨ Ø.26 82°
AFTER KNURLING

Ⓐ

.50

14P
DIAMOND
KNURL

Ø.8750
+.0002
–.0000

Ⓔ

.50

NECK .12 × R.12
⊔ .12

3.500±.001

Ⓐ

Ⓒ

0.125:1

G

Ⓓ

NECK .12 × R .12 ⊔ .01

.250 × .125 KEYSEAT

.75

1.12

Ⓐ

R .12

Ø1.250
Ø1.249

5/8 – 11 UNC – ⊔ .88 – LH

NO.	CHANGE	DATE	BY	UNLESS OTHERWISE SPECIFIED TOLERANCES ON DIMENSIONS ARE:	
Ⓐ	WAS 1.14	5-3-08	T.P.O.	DECIMAL DIMENSIONS ±.01" (TWO-PLACE) ±.002" (THREE-PLACE)	
Ⓐ	.125 ADDED	5-3-07	T.P.O.	ANGULAR DIMENSIONS ±.45°	
NO.	CHANGE	DATE	BY	**CUTTER ADAPTER**	**BP-22**

ASSIGNMENT—UNIT 22: CUTTER ADAPTER (BP-22)

Student's Name _____

1. What tolerances are given for:

   (a) Angular dimensions?

   (b) Decimal dimensions?

2. Give the body taper specifications.

3. What is the taper per inch for the body?

4. What is the diameter at the small end of the outside taper?

5. Determine basic dimension (A).

6. Give the taper for the hole.

7. What is the taper per inch for the hole?

8. Give the diameter at the small end of the inside taper.

9. Compute basic dimension (B).

10. Give the outside thread specifications.

11. What does the LH in the thread note for the shank end specify?

12. What thread class is the 2″ threaded nose?

13. What operation is performed on the 2.75″ diameter?

14. How is this operation specified?

15. Determine the maximum diameter for the reamed hole.

16. Give the note for the four holes in the largest diameter.

17. Specify the chamfer for the knurled area.

18. Specify the angle of chamfer for the 2″ threaded nose.

19. How deep is the 2″ threaded chamfer?

20. Give the original basic length of the tapered body before any change was made.

21. Compute the maximum length of dimension (C).

22. What matching operation is indicated by the finish symbol ∘/ ?

23. Determine basic dimension (D).

24. Give the dimensions of the keyseat.

25. Determine the basic overall length of the adapter.

26. Name three different types of fits.

27. Describe briefly the purpose of (a) positive clearance and (b) negative clearance.

28. Give the limit dimensions for the .8750 hole.

1. (a) _____

   (b) _____

2. _____

3. _____

4. _____

5. (A) _____

6. _____

7. _____

8. _____

   _____

9. (B) _____

10. _____

11. _____

12. _____

13. _____

14. _____

    _____

15. _____

16. _____

    _____

17. _____

18. _____

19. _____

20. _____

21. (C) _____

22. _____

23. (D) _____

24. Width _____

    Depth _____

25. _____

26. (a) _____

    (b) _____

    (c) _____

27. (a) _____

    _____

    _____

    (b) _____

    _____

    _____

28. _____ _____

The SI Metric System

Metric System Dimensioning and ISO Symbols

The International System of Units (SI) was established by agreement among many European nations to provide a logical interconnected framework for all measurements used in industry, science, and commerce. SI is a modernized, accepted adaptation of many European metric systems. SI metrics relates to seven basic units of measurement: (1) length, (2) time, (3) mass, (4) temperature, (5) electric current, (6) luminous intensity and (7) amount of substance. Multiples and submultiples of these basic units are expressed in decimals. All additional SI units are derived from the seven basic units.

THE SI METRIC SYSTEM OF MEASUREMENT

Prefixes are used in the metric system to show how the dimension relates to a basic unit. For example, deci- means one-tenth of the basic unit of measure. Thus, 1 decimeter = 0.1 of a meter; 1 deciliter = 0.1 liter; and 1 decigram = 0.1 gram. Similarly, centi- = one hundredth; and milli- = one thousandth.

Dimensions larger than the basic unit of measure are expressed with the following prefixes: deka- = ten times greater; hecto- = one hundred times greater; and kilo- = 1000 times greater. To repeat, the most common unit of metric measure used on engineering drawings is the millimeter (mm). The next frequently used unit is the meter (m). Surface finishes are given in micrometers (millionths of a meter), using the symbol μm.

WORLDWIDE DRAFTING STANDARDS

Two sets of drafting standards are accepted worldwide: ANSI (inch-metric) and ISO (metric). Almost all practices (dimensioning; line types and thicknesses; welding symbols; drawing sheet sizes; GD&T; screw threads; surface textures; finish symbols; etc.) are now the same in either system or roughly comparable. Most major technical differences are beyond the scope of this text.

Of course, the most obvious difference between the two is that U.S. multi-view drawings use third-angle projection. Europe and Asia follow ISO practices and view objects from first-angle projection (see Unit 24).

ANSI recommends decimal-inch dimensioning and notes on U.S. drawings to be unidirectional. ISO default practice allows aligned decimal millimeter dimensions. Both standards are used throughout North America.

CONVERTING METRIC AND CUSTOMARY INCH SYSTEM DIMENSIONS

Converting measurements from SI millimeter units to U.S. customary inch units (CIU) or vice versa on a drawing can be done by several CAD programs (Unit 30) or by using handbooks, Figure 23–1.

Finding equivalent dimensional values in the other system can be done mathematically by using hard conversions or by making soft conversions.

Trade and engineering handbooks contain conversion charts that simplify the process of determining in one system the equivalent value of a dimension given in the other system. Sections of a conversion table are illustrated in Figure 23–1 to show customary inch units and equivalent metric units. Figure 23–1 also includes a drill series ranging from #80 to #1, letter size drills from (A) to (Z), and fractional dimensions from .001″ to 1.000″. Note that millimeter equivalents, correct to four decimal places, are shown for both the drill sizes and the decimals, which represent fractional parts of an inch.

The millimeter range of the table is from 0.0254 mm to 25.4000 mm, corresponding to 0.001″ to 1.000″, respectively. The millimeter equivalent of a measurement in the inch system may be found by multiplying the decimal value by 25.40.

Drill No. or Letter	Inch Standard		ISO Metric Standard	Drill No. or Letter	Inch Standard		ISO Metric Standard
	Fractional or Decimal Size		mm Size		Fractional or Decimal Size		mm Size
		.001	0.0254			.401	10.1854
		.002	0.0508			.402	10.2108
		.003	0.0762			.403	10.2362
		.004	0.1016	Y	13/32	.404	10.2616
		.005	0.1270			.405	10.2870
		.006	0.1524			.406	10.3124
		.007	0.1778		13/32	.4062	10.3187
		.008	0.2032			.407	10.3378
		.009	0.2286			.408	10.3632
		.010	0.2540			.409	10.3886
		.011	0.2794				
		.012	0.3048				
80	.0135	.013	0.3302				
79	.0145	.014	0.3556			.996	25.2984
		.015	0.3810			.997	25.3238
	1/64	.0156	0.3969			.998	25.3492
78		.016	0.4064			.999	25.3746
		.017	0.4318			1.000	25.4000

FIGURE 23–1
Portions of a conversion table showing drill sizes and millimeter equivalents to inch dimensions.

ROUNDING OFF LINEAR DIMENSIONS

Many dimensions that are converted from inches to millimeters or from millimeters to inches are rounded off. With 1 mm = .03937″, conversions can be made to very high limits of accuracy. This is done by calculating the decimal value to a greater number of digits than is required within the range of tolerances.

Production costs are related directly to the degree of accuracy required to produce a part. Therefore, the limits of dimensions should be rounded off to the least number of digits in the decimal dimension that will provide the greatest tolerance and ensure interchangeability.

A decimal dimension may be rounded off by increasing the last required digit by 1 if the digit that follows on the right is 5 or greater or by leaving the last digit unchanged if the digit to the right is less than 5. For example:

1.5875″	rounded off to three decimal places =	1.588″
1.5874 mm	rounded off to three decimal places =	1.587 mm
1.646 mm	rounded off to two decimal places =	1.65 mm
1.644″	rounded off to two decimal places =	1.64″

When dimensions on a drawing need to be divided, the part designer tries to round off the dimensions to the nearest even decimal. This practice permits parts to be economically machined to tolerances within a required number of decimal places.

PROJECTION SYMBOLS

The International Organization for Standardization (ISO) recommends the use of a projection symbol on drawings that are produced in one country for use among many countries, Figure 23–2. The projection symbols are intended to promote the accurate exchange of technical information through drawings.

The ISO projection symbol, the notation on tolerances, and information on whether metric and/or inch dimensions are used on the drawing should appear as notes either within the title block or adjacent to it, Figure 23–3. The designation THIRD-ANGLE PROJECTION is not always included with the symbol on a drawing.

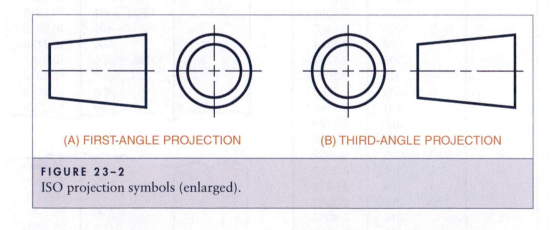

(A) FIRST-ANGLE PROJECTION   (B) THIRD-ANGLE PROJECTION

FIGURE 23–2
ISO projection symbols (enlarged).

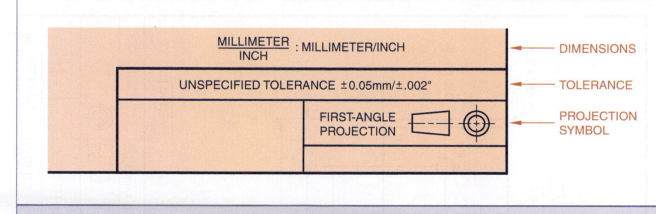

FIGURE 23-3
Dimension, tolerance, and symbol notes in a title block.

UNILATERAL AND BILATERAL MILLIMETER TOLERANCING

The practice for **unilateral millimeter tolerancing** on ISO metric drawings (where either the + or − tolerance is zero) is to use a single zero without a plus + or minus − sign. For example, if a positive + millimeter tolerance of 0.25 and a zero negative tolerance applies to a 50 mm dimension, the toleranced dimension on a SI metric drawing is $50^{+0.25}_{0}$. If a −0.25 mm and zero positive tolerance is required, the toleranced dimension is given as $50^{0}_{-0.25}$.

A bilateral tolerance such as +0.75 mm and −0.1 mm, applied to a 125 mm dimension, appears on a drawing as $125^{+0.75}_{-0.10}$. Note that one or more zeros are added so both the plus + and − values have the same number of decimal places.

Where limit dimensioning is used for millimeters, the maximum or minimum values have digits following the decimal point; the other value will then have zeros added for uniformity. An example would be 25.45 UL shown as 25.00 LL. An incorrect version would be 25.45 UL shown simply as 25 LL with no zeros.

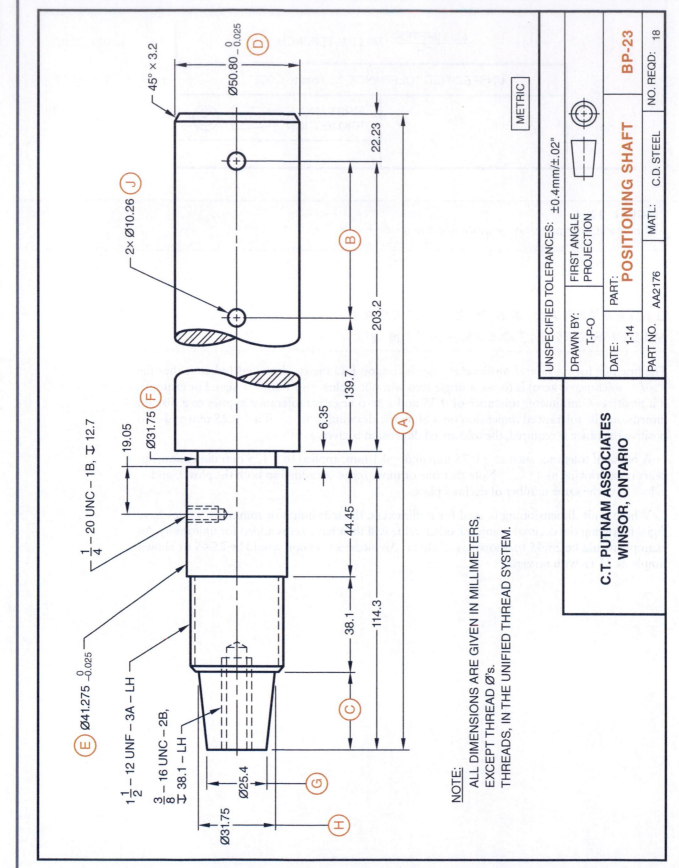

45° × 3.2

Ø50.80 −0.025/0

D

2× Ø10.26 J

22.23

B

203.2

A

139.7

6.35

F

Ø31.75

19.05

¼ – 20 UNC – 1B, ⊥ 12.7

E Ø41.275 −0.025/0

1½ – 12 UNF – 3A – LH

⅜ – 16 UNC – 2B,
⊥ 38.1 – LH

44.45

38.1

114.3

C

Ø25.4

G

Ø31.75

H

NOTE:
ALL DIMENSIONS ARE GIVEN IN MILLIMETERS,
EXCEPT THREAD Ø's.
THREADS, IN THE UNIFIED THREAD SYSTEM.

UNSPECIFIED TOLERANCES: ±0.4mm/±.02″		
DRAWN BY: T-P-O	FIRST ANGLE PROJECTION	
DATE: 1-14	PART: **POSITIONING SHAFT**	
PART NO. AA2176	MATL: C.D. STEEL	

METRIC

C.T. PUTNAM ASSOCIATES
WINSOR, ONTARIO

BP-23

NO. REQD: 18

ASSIGNMENT—UNIT 23: POSITIONING SHAFT (BP-23)

Student's Name _____

1. State what system of projection is used.
2. Indicate what unit of measurement is used for dimensioning.
3. Give the tolerance that applies on (a) metric dimensions and (b) dimensions converted to the inch system, when no tolerance is specified.
4. Compute the basic metric dimension for overall length Ⓐ rounded off to two decimal places.
5. Determine the basic center-to-center distance between the two holes Ⓑ, in millimeters.
6. Find the basic length of the tapered portion Ⓒ.
7. Determine the equivalent customary (inch) units for each linear dimension, rounded off to two decimal places. Use a four-place millimeter/inch conversion chart, if available.
8. Convert each metric diametral dimension to its equivalent inch dimension, correct to three decimal places.
9. Give the letter size drill that corresponds to the drill size of the two Ⓙ holes.
10. State what system of representation is used on the threaded areas.
11. Convert the depth of each tapped hole to its equivalent in the inch system.
12. Tell what each part of the three thread dimensions designates.
13. Give the minimum and maximum limits of diametral dimensions Ⓓ and Ⓔ in metric and inch dimensions, correct to three decimal places.
14. Determine the minimum and maximum limits of diametral dimensions Ⓕ, Ⓖ, and Ⓗ in metric and inch dimensions, rounded to two decimal places.
15. Compute the lower and upper dimensional limits of each linear dimension in both the metric and inch systems. Round off the dimensions to two decimal places.

1. _____
2. _____
3. (a) _____
   (b) _____
4. Ⓐ _____   5. Ⓑ _____
6. Ⓒ _____

7.

Millimeters	Inch Equiv.	Millimeters	Inch Equiv.	Millimeters	Inch Equiv.
=		=		=	
=		=		=	
=		=		=	
=		=		=	

8.

Millimeters	Inch Equiv.
=	
=	
=	

9. _____

10. _____
11. _____
12. _____

13.

	Millimeters	Inch Equiv.
Ⓓ Min.		
Max.		
Ⓔ Min.		
Max.		

14.

	Millimeters	Inch Equiv.
Ⓕ Min.		
Max.		
Ⓖ Min.		
Max.		
Ⓗ Min.		
Max.		

15.

Dimension	Metric (mm)		Inch	
	Min.	Max.	Min.	Max.

24. First-Angle Projection and Dimensioning

PROJECTION BASED ON SYSTEM OF QUADRANTS

The position each view of an object occupies, as treated thus far in this text, is based on the U.S. and Canadian standard of third-angle projection. Third-angle projection is derived from a theoretical division of all space into four quadrants. The horizontal plane in Figure 24–1 represents an X-axis. The vertical plane is the Y-axis. The four quadrants produced by the two planes are shown as I, II, III, and IV.

It is possible to place an object in any one of the four quadrants. Views of the object may then be projected. The third quadrant was adopted in the United States and Canada because the projected views of an object occupy a natural position. Drawings produced by such standards of projection are comparatively easy to interpret. Each view is projected so the object is represented as it is seen. The principles of third-angle projection are reviewed graphically in Figure 24–2.

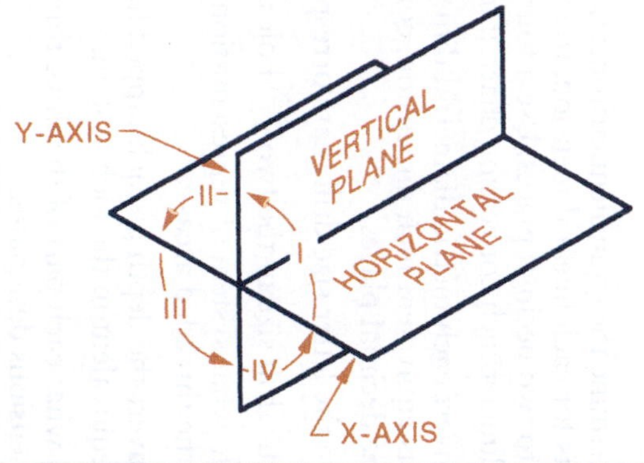

METRIC: FIRST-ANGLE PROJECTION

Increasing worldwide commerce and the interchange of materials, instruments, machine tools, and other precision-made parts and mechanisms require the interpretation of drawings that have been prepared according to different systems of projection. The accepted standard of projection of the Common Market and other countries of the world relates to the first quadrant (I). The system, therefore, is identified as first-angle projection.

FIGURE 24–1
Four quadrants.

Dimensions given in the metric system are used with first-angle drawings. As world leaders in business and industry, the United States and Canada are also using a great number of first-angle projection drawings.

The quadrants and X and Y axes in first-angle projection are shown in Figure 24–3. Arrows are used in Figure 24–3A to illustrate positions from which the object in quadrant I may be viewed. The three common views are named front view, top view, and left-side view. These views are positioned in Figure 24–3B as they would appear (without the imaginary projection box outline) on a drawing.

In contrast with third-angle projection, the top view in first-angle projection appears under the front view. Similarly, the left-side view is drawn in the position occupied by the right-side view in third-angle projection. This positioning, which is not natural, makes it more difficult to visualize and interpret first-angle drawings. Each view in first-angle projection is projected through the object from a surface of the object to the corresponding projection plane. However, the views in

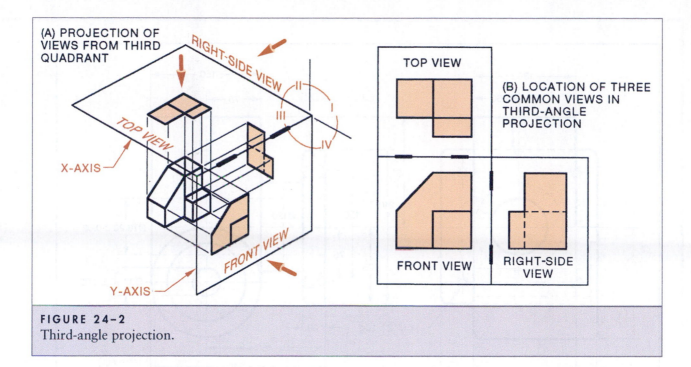

FIGURE 24–2
Third-angle projection.

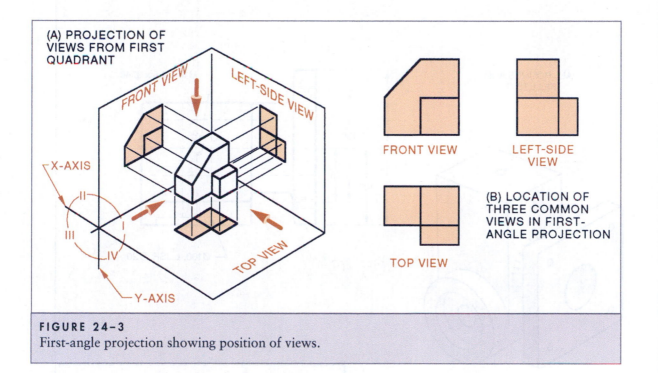

FIGURE 24–3
First-angle projection showing position of views.

both first-angle and third-angle projection provide essential information for the craftsperson to produce or assemble a single part or a complete mechanism.

DIMENSIONING FIRST-ANGLE DRAWINGS

Dimensions on first-angle drawings are in metric units of measure. When a numerical value is less than 1, the first numeral is preceded by a zero and a decimal point. For example, the decimal .733 is dimensioned to read 0.733. Whole numbers do not require a zero to the right of the decimal point. Twenty-four (24 mm) would not be written as 24.0; it would be 24 mm.

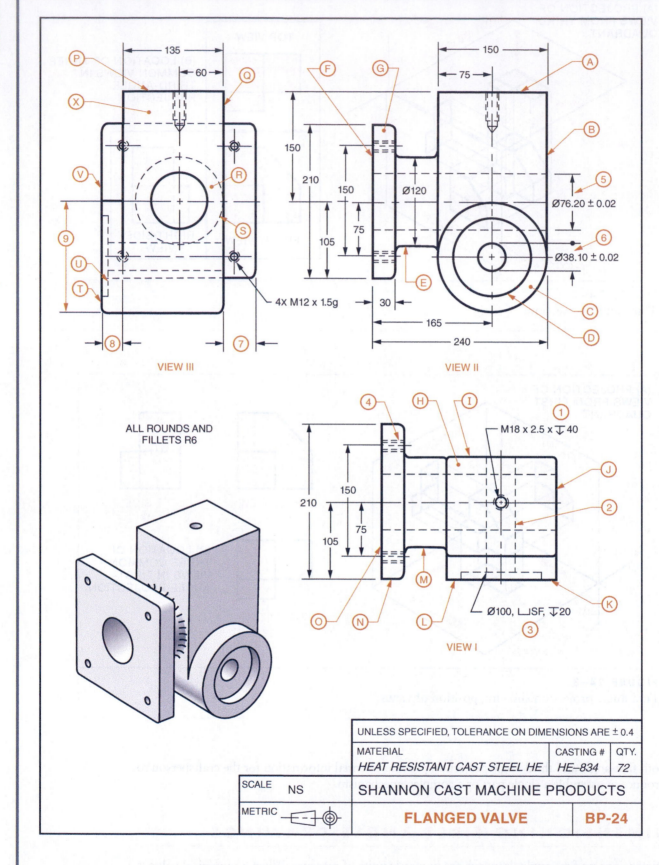

135
60

ⓟ
ⓧ
ⓠ
ⓥ
ⓡ
ⓢ
9
ⓤ
ⓣ
8
7

4x M12 x 1.5g

VIEW III

ⓕ
ⓖ
ⓐ
150
75
150
210
150
Ø120
75
ⓑ
5
Ø76.20 ± 0.02
6
Ø38.10 ± 0.02
105
ⓒ
ⓔ
30
165
ⓓ
240

VIEW II

ALL ROUNDS AND
FILLETS R6

4
ⓗ
ⓘ
1
M18 x 2.5 x ⌵ 40
ⓙ
210
150
75
2
105
ⓜ
ⓚ
Ø100, ⌴SF, ⌵20
ⓞ
ⓝ
ⓛ
3

VIEW I

UNLESS SPECIFIED, TOLERANCE ON DIMENSIONS ARE ± 0.4		
MATERIAL	CASTING #	QTY.
HEAT RESISTANT CAST STEEL HE	*HE–834*	*72*
SCALE NS	SHANNON CAST MACHINE PRODUCTS	
METRIC	**FLANGED VALVE**	**BP-24**

ASSIGNMENT—UNIT 24: FLANGED VALVE (BP-24)

Student's Name _____

1. (a) Indicate the material used to manufacture the **flanged valve**.

   (b) Identify the casting number and the required quantity.

2. (a) Name the system of projection.

   (b) State the name of each view.

   (c) Identify the system and the unit of measurement used on the drawing.

3. Locate surfaces (A), (B), and (C), in **VIEW I** and **VIEW III**.

4. Identify features (K), (M), and (N) in **VIEW II** and **VIEW III**.

5. Give the letter(s) and number(s) in **VIEW I** that relate to the following features:

   (a) Flange

   (b) Flange hub

   (c) Flange screw threads

   (d) Threads in block

   (e) 38.1 hole

6. Determine the basic overall machined (a) width, (b) height, and (c) depth of the part.

7. Write the specification of the tapped hole.

8. (a) Determine the maximum width, height, and thickness of the flange.

   (b) Give the size of the corner rounds and fillets.

9. Establish the minimum and maximum center-to-center distances (along X and Y axis) between the tapped holes in the flange plate.

10. (a) Calculate the minimum and maximum diameter of holes (5) and (6).

    (b) Give the basic length of holes (5) and (6).

11. Calculate the basic following dimensions:

    (a) (7)

    (b) (8)

    (c) (9)

1. (a) _____

   (b) Casting # _____

   Quantity _____

2. (a) _____

   (b) VIEW I _____

   VIEW II _____

   VIEW III _____

   (c) _____

3.   VIEW I   VIEW III

   (A) = _____   _____

   (B) = _____   _____

   (C) = _____   _____

4.   VIEW II   VIEW III

   (K) = _____   _____

   (M) = _____   _____

   (N) = _____   _____

5.   Identification
                (VIEW I)

   (a) Flange _____

   (b) Flange hub _____

   (c) Flange screw threads _____

   (d) Threads in block _____

   (e) 38.1 hole _____

6. (a) Width _____

   (b) Height _____

   (c) Depth _____

7. _____

8. (a) Maximum:

   Width _____

   Height _____

   Thickness _____

   (b) Rounds and Fillets _____

9. (a) Minimum _____

   (b) Maximum _____

10. (a) (5) Min. _____ Max. _____

    (6) Min. _____ Max. _____

    (b) (5) _____

    (6) _____

11. (7) = _____

    (8) = _____

    (9) = _____

25

Metric Screw Threads, Dual Dimensioning, and Tolerancing

Over time, one standard has evolved for metric threads, replacing many thread variations that have existed in European countries. This standard is the ISO metric M thread series, which is similar to the Unified form of threads covered in Unit 19.

THREAD NOTES: BASIC THREAD AND TOLERANCE CALLOUTS

Thread notes provide complementary information to that given by graphically representing and dimensioning screw threads on a drawing. Dimensions and other data contained in three notes for SI metric screw threads follow ISO standards. The necessary information for producing general-purpose metric screw threads is arranged in a standardized sequence in what is called a basic metric thread callout. The callout is referenced to the thread by using a dimensioning leader. Figure 25–1 identifies the sequencing of data for ISO metric screw threads.

Figure 25–2 shows the same metric thread with tolerance callout data added. Note the length (32 mm) of thread may be given on either the thread drawing or in the thread callout.

Both metric threads and Unified threads can be drawn in pictorial, schematic, or simplified representation (19–5) and dimensioned on drawings according to either Unified screw

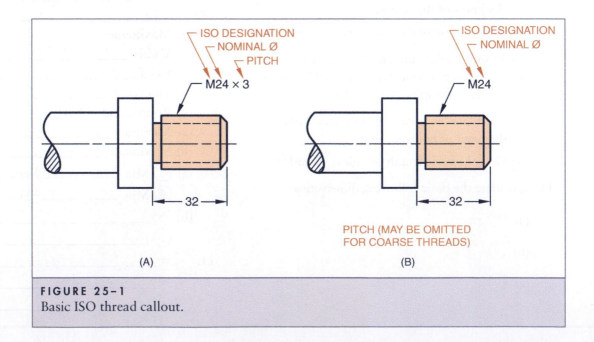

(A)   (B)

FIGURE 25–1
Basic ISO thread callout.

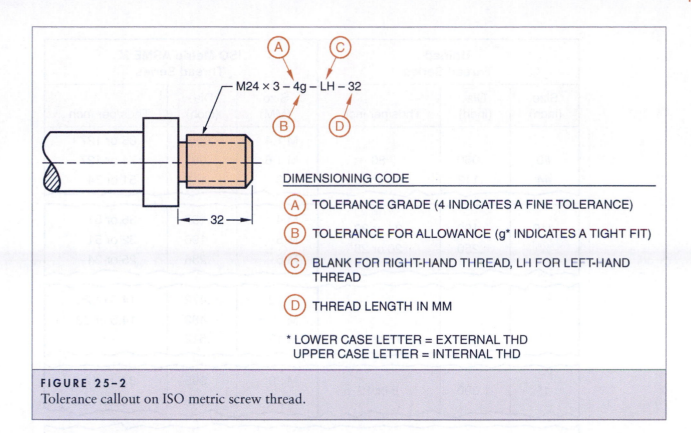

FIGURE 25-2
Tolerance callout on ISO metric screw thread.

thread or ISO metric specifications. The general range of ISO metric threads is M1 × 0.25 to M 300 × 6.

Screw thread conversion tables contain engineering data on major (basic) outside diameters and pitch for ISO metric screw threads, Unified thread series sizes, and the best ISO metric equivalents for selected American screw thread sizes. For example, a 1/2 (.500) − 20 UNF designation on a drawing indicates an outside thread diameter of .500″, a pitch (distance between two identical points on successive threads) of 20 threads per inch, and a Unified Fine Thread Series. Although currently there is no precise metric equivalent, the outside diameter and pitch (expressed in metric units) are 12.5 mm and 1.25 mm respectively. This example may be dual dimensioned as shown in Figure 25–3.

ISO metric and **ASME M** profile thread series show the same basic profile as the Unified inch series threads. ISO metric threads are interchangeable with ASME M profile threads. However, they are not interchangeable with Unified thread series because of size variations. Figure 25–4 shows a sample of these differences.

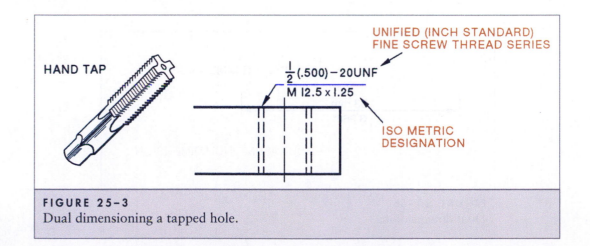

FIGURE 25-3
Dual dimensioning a tapped hole.

Unified Thread Series			ISO Metric ASME M Thread Series		
Size (inch)	Dia. (inch)	Thds/per inch	Size (MM)	Dia. (inch)	Thds/per inch
#0	.060	80	M 1.4	.055	85 or 127
			M 1.6	.063	74 or 127
#4	.112	40 or 48	M 3	.118	51 or 74
#10	.190	24 or 32	M 4	.157	36 or 51
			M 5	.196	32 or 51
¼"	.250	20 or 28	M 6	.236	25 or 34
			M 12	.472	14.5 or 20
			M 12.5*	.492	14.5 or 20
½"	.500	13 or 20	M 13	.512	
1"	1.000	8 or 12	M 20	.995	8.5 or 12.5
			M 27	1.063	8.5 or 12.5

*NONPREFERRED SIZE

FIGURE 25–4
Selected comparison of unified threads and the closest metric thread size.

DUAL DIMENSIONING: CUSTOMARY INCH AND SI METRIC

Dual dimensioned parts may be produced in any country regardless of whether Customary inch system dimensions or ISO metric dimensions, or both, are used. Dual dimensioning implies that each dimension is given both in metric and English units. The metric dimension (including tolerances) is given on one side of the dimension line; the inch dimension is given on the opposite side of the line. For example in Figure 25–5a, if dual dimensioning is used on a dimension of 83 millimeters (3.268"), the dimension is shown as:

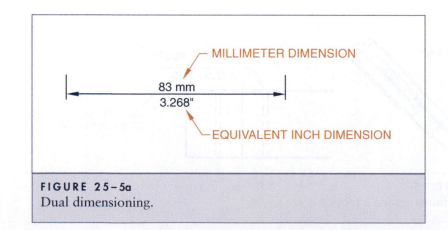

FIGURE 25–5a
Dual dimensioning.

If a tolerance of ±0.2 mm applies to the base dimension of 83 mm, the dimension and tolerance in the Customary inch system equals 3.268″ ± .008″. The dual dimension (as shown in Figure 25–5b) is represented as:

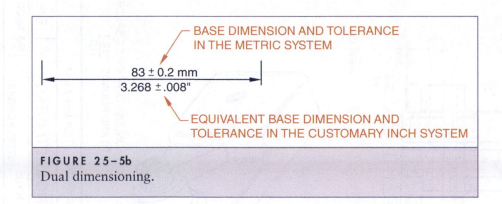

FIGURE 25–5b
Dual dimensioning.

Instead of a dimension line between the dimensions in the two systems, some industries use the slash (/) symbol between the metric and Customary inch units for dual dimensioning as shown in Figure 25–5c.

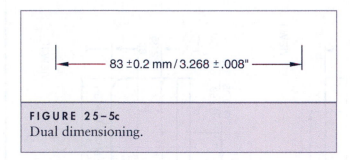

FIGURE 25–5c
Dual dimensioning.

UNILATERAL AND EQUAL/UNEQUAL BILATERAL TOLERANCES

The previous examples showed equal tolerances in both directions (±). However, some tolerances may be unequal or unilateral. If unequal bilateral tolerances of $^{+0.2}_{-0.1}$ mm are applied to the 83 mm dimension, the drawing (with the scale and dimensional systems indicated) is dimensioned as (Figure 25–5d):

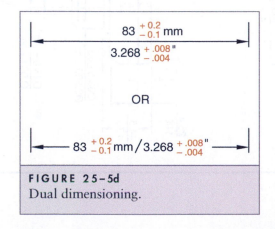

FIGURE 25–5d
Dual dimensioning.

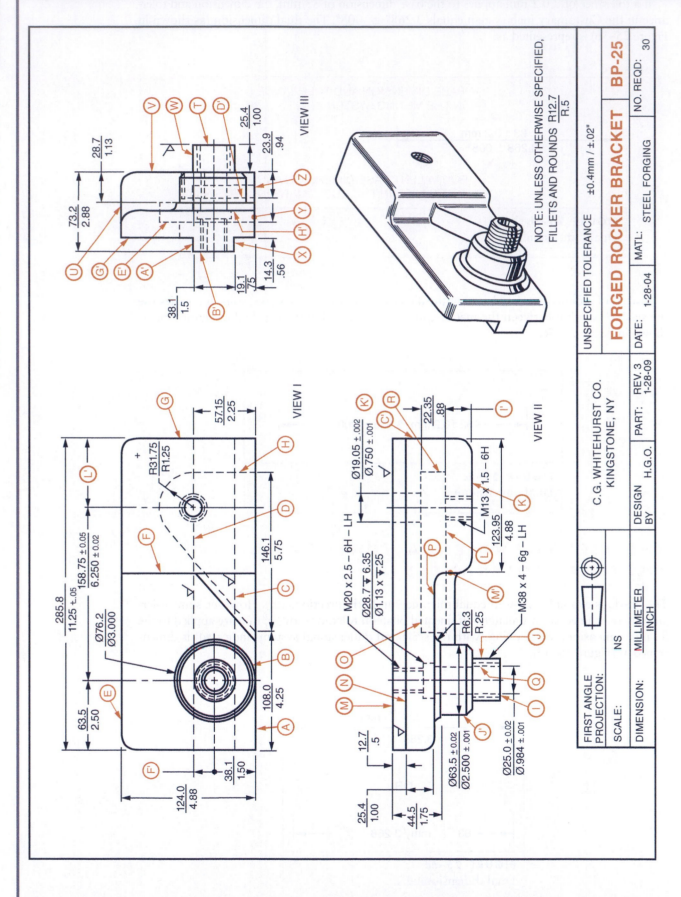

VIEW III

VIEW I

VIEW II

NOTE: UNLESS OTHERWISE SPECIFIED,
FILLETS AND ROUNDS R12.7 / R.5

UNSPECIFIED TOLERANCE	±0.4mm / ±.02"

FORGED ROCKER BRACKET **BP-25**

MATL: STEEL FORGING	NO. REQD: 30

DATE:	1-28-04

C.G. WHITEHURST CO.
KINGSTONE, NY

| PART: | REV. 3 |
| | 1-28-09 |

DESIGN BY	H.G.O.

FIRST ANGLE PROJECTION:	
SCALE:	NS
DIMENSION:	MILLIMETER
	INCH

ASSIGNMENT—UNIT 25: FORGED ROCKER BRACKET (BP-25)

Student's Name _____

1. Name VIEWS I, II, and III.

2. Name the dimensions above the line if these represent the product design dimensions.

3. State what dual dimensioning implies.

4. Determine what system of tolerancing is used.

5. Locate surfaces Ⓐ, Ⓒ, Ⓓ, and Ⓔ in VIEW III. Give the corresponding letter.

6. Give the letters in VIEW II that identify surfaces Ⓕ, Ⓖ, and Ⓗ.

7. Locate surfaces Ⓘ, Ⓙ, Ⓚ, Ⓛ, Ⓜ, Ⓝ, Ⓞ, and Ⓟ in VIEW III.

8. Determine the basic distance between the center of the .750″ hole and surface Ⓖ, correct to two decimal places. Give the dimension in metric and customary inch units.

9. Give the center distance Ⓕ′ in millimeters and inches, correct to two decimal places.

10. Locate thickness Ⓘ′ in both dimensioning systems, correct to two decimal places.

11. Indicate the tolerance to use if none is specified.

12. Interpret the meaning of the screw thread designation M38×4-6g-LH.

13. Give the outside diameter and pitch (in millimeters) for each metric screw thread.

14. Convert each metric screw thread to its closest equivalent American screw thread.

15. Give the basic diameter of Ⓙ and the lower dimensional limit of Ⓙ′ in inch units, to three decimal places.

16. Determine the maximum metric dimensions for Ⓚ′ and Ⓛ′, correct to two decimal places.

1. _____

2. _____

3. _____

4. _____

5. Ⓐ = _____ Ⓓ = _____
   Ⓒ = _____ Ⓔ = _____

6. Ⓕ = _____ Ⓖ = _____
   Ⓗ = _____

7. Ⓘ = _____ Ⓜ = _____
   Ⓙ = _____ Ⓝ = _____
   Ⓚ = _____ Ⓞ = _____
   Ⓛ = _____ Ⓟ = _____

8. _____

9. _____

10. _____

11. _____

12. _____

13. _____

14. _____

15. Ⓙ = _____ Ⓙ′ = _____

16. Ⓚ′ = _____ Ⓛ′ = _____

Sections

26 Cutting Planes, Full Sections, and Section Lining

An exterior view shows the object as it looks when seen from the outside. The inside details of such an object are shown on the drawing by hidden lines.

As the details inside the part become more complex, additional invisible lines are needed to show the hidden details accurately. This tends to make the drawing increasingly more difficult to interpret. One technique used on such drawings to simplify them is to cut away a portion of the object, resulting in a sectional view. This exposes the inside surfaces. On cutaway sections, all of the edges that are visible are represented by visible edge or object lines.

To obtain a sectional view, an imaginary cutting plane is passed through the object as shown in Figure 26–1A. Figure 26–1B shows the front portion of the object removed. The direction and surface through which the cutting plane passes is represented on the drawing by a cutting plane line. The exposed surfaces, which have been theoretically cut through, are further identified by a number of fine slant lines called section or cross-hatch lines. If the cutting plane passes completely through the object, the sectional view is called a full section. A section view is often used in place of a regular view. Hidden lines are normally omitted from sectional views unless they add to clarity.

CUTTING PLANE LINES

The cutting plane line is either a thick (heavy) line with one long and two short dashes or a series of thick (heavy), equally spaced long dashes. Examples are shown in Figure 26–2. The line represents the edge of the cutting plane. The arrowheads on the ends of the cutting plane line show the direction in which the section is viewed. A capital letter is usually placed at the end of the cutting plane line to identify the section. Cutting plane lines can be drawn in simplified form, Figure 26–2C. At times, section view(s) must be removed and located away from a regular view(s), Figure 26–2D.

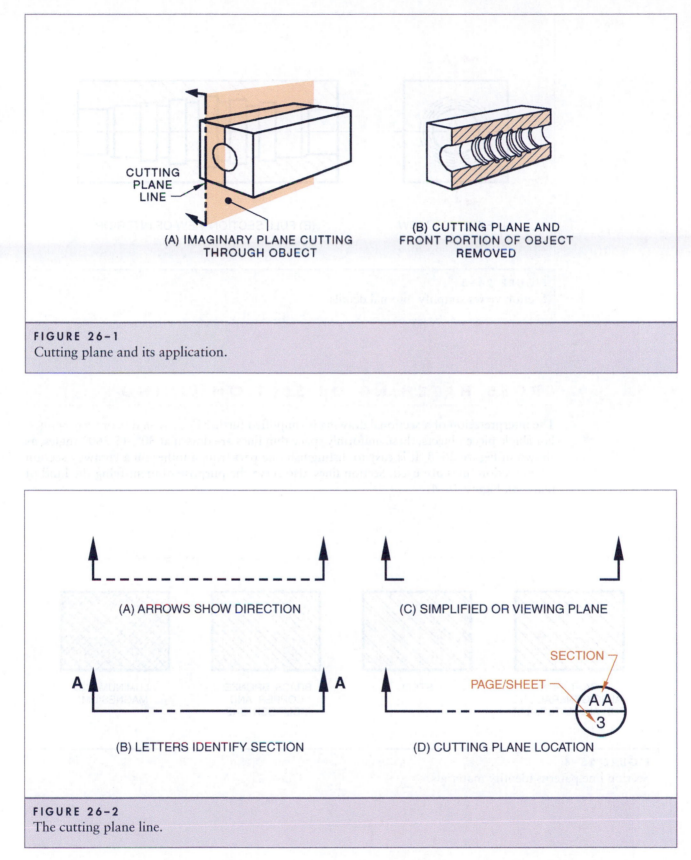

FIGURE 26–1
Cutting plane and its application.

(A) IMAGINARY PLANE CUTTING THROUGH OBJECT

CUTTING PLANE LINE

(B) CUTTING PLANE AND FRONT PORTION OF OBJECT REMOVED

(A) ARROWS SHOW DIRECTION

(C) SIMPLIFIED OR VIEWING PLANE

(B) LETTERS IDENTIFY SECTION

A   A

(D) CUTTING PLANE LOCATION

SECTION
PAGE/SHEET
A A
3

FIGURE 26–2
The cutting plane line.

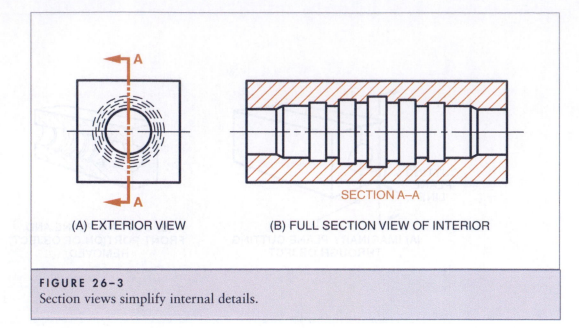

(A) EXTERIOR VIEW   (B) FULL SECTION VIEW OF INTERIOR

SECTION A–A

FIGURE 26-3
Section views simplify internal details.

CROSS-HATCHING OR SECTION LINING

The interpretation of a sectional drawing is simplified further by cross-hatch or section lines. For single piece objects, these uniformly space thin lines are drawn at 30°, 45°, 60° angles, as shown in Figure 26–3. It is easy to distinguish one part from another on a cutaway section when section lines are used. Section lines also serve the purpose of identifying the kind of material, Figure 26–4.

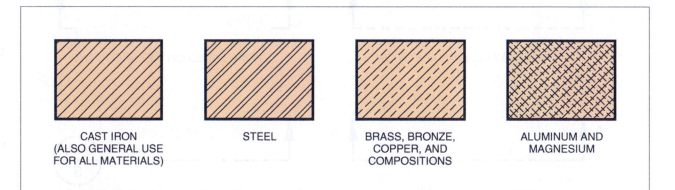

CAST IRON
(ALSO GENERAL USE
FOR ALL MATERIALS)

STEEL

BRASS, BRONZE,
COPPER, AND
COMPOSITIONS

ALUMINUM AND
MAGNESIUM

FIGURE 26-4
Section line paterns identify materials.

REVOLVED AND REMOVED SECTIONS

In order to show clearly the true shape of elongated objects, such as ribs, bars, arms, spokes, and webs, Figure 26–5A, revolved and removed sections are used.

A revolved section view uses a centerline instead of a cutting plane to indicate where a section has been taken and rotated 90°, Figure 26–5B.

Another technique used to show the shape and size of a specific area of an object is a removed section. This method, Figure 26–5C, uses a cutting plane showing where a section has been taken, removed from the view, labeled, and relocated on a drawing sheet of the object. If necessary, it can be enlarged and dimensioned.

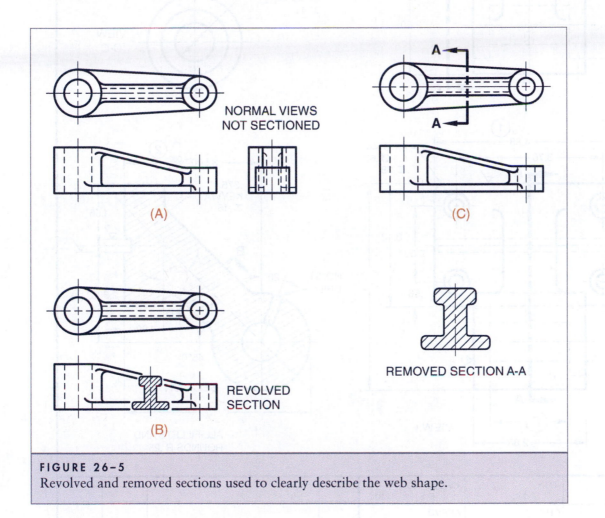

FIGURE 26–5
Revolved and removed sections used to clearly describe the web shape.

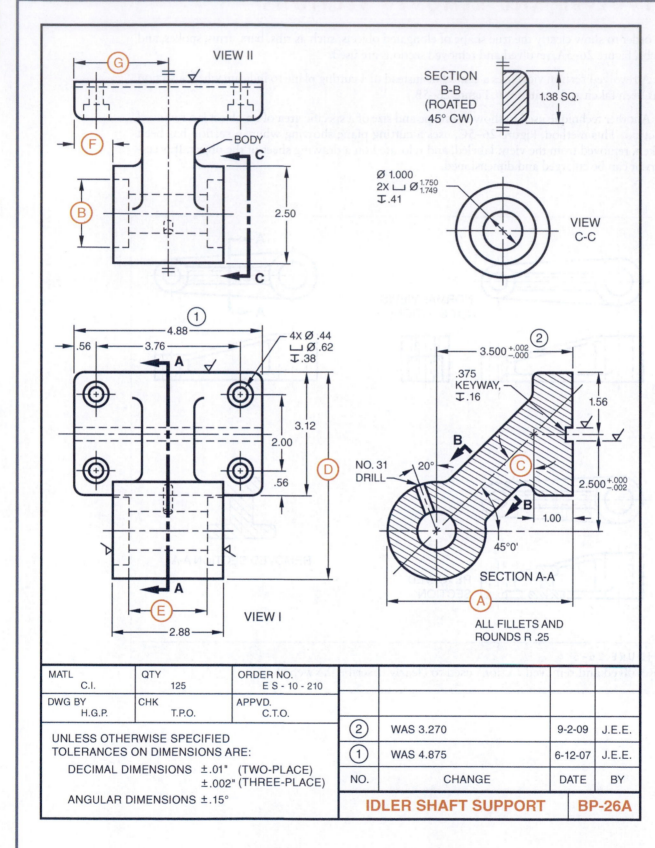

VIEW II

G

F

B

BODY

SECTION
B-B
(ROATED
45° CW)

1.38 SQ.

C
C

2.50

Ø 1.000
2X ⌴ Ø 1.750 / 1.749
⌆ .41

VIEW
C-C

1

4.88

.56   3.76

A

4X Ø .44
⌴ Ø .62
⌆ .38

3.12

2.00

.56

D

E

VIEW I

2.88

2

3.500 +.002 / -.000

.375
KEYWAY,
⌆ .16

1.56

B

NO. 31
DRILL

20°

C

2.500 +.000 / -.002

B

1.00

45°0'

SECTION A-A

A

ALL FILLETS AND
ROUNDS R .25

MATL C.I.	QTY 125	ORDER NO. E S - 10 - 210			
DWG BY H.G.P.	CHK T.P.O.	APPVD. C.T.O.			
UNLESS OTHERWISE SPECIFIED TOLERANCES ON DIMENSIONS ARE:			2	WAS 3.270	9-2-09	J.E.E.

2	WAS 3.270	9-2-09	J.E.E.
1	WAS 4.875	6-12-07	J.E.E.
NO.	CHANGE	DATE	BY

UNLESS OTHERWISE SPECIFIED
TOLERANCES ON DIMENSIONS ARE:

DECIMAL DIMENSIONS ±.01" (TWO-PLACE)
±.002" (THREE-PLACE)

ANGULAR DIMENSIONS ±.15°

IDLER SHAFT SUPPORT   BP-26A

ASSIGNMENT—UNIT 26A: IDLER SHAFT SUPPORT (BP-26A)

Student's Name _____

1. What is the name of line **C-C**?

2. What is the purpose of line **C-C**?

3. What two types of sections are shown in **B-B**?

4. Give the specifications for the flat keyway.

5. What size are the fillets and rounds?

6. Compute angle Ⓒ from dimensions given on the drawing.

7. Give the maximum diameter for the 1" hole.

8. What is the upper-limit of Ⓑ?

9. How are the machined surfaces indicated?

10. What type line shows where Section **A-A** is taken? In what view?

11. Determine basic dimensions Ⓐ and Ⓓ.

12. Determine from Section **B-B** what material is required.

13. How wide and thick is the **body?**

14. Indicate what two changes were made from the original drawing.

15. Determine basic dimension Ⓔ.

16. Compute UL dimension Ⓕ.

17. What is the upper-limit of dimension Ⓖ?

18. Why is section **A-A** a full section?

19. Show the section linings for aluminum and for steel.

20. Identify (a) the system of dimensioning and (b) the classification of the unspecified tolerances.

21. Change the tolerances as follows: two-place decimal dimensions $\pm^{.02''}_{.00''}$ three-place $\pm^{.003''}_{.000''}$ and angular $\pm^{1.5°}_{.00°}$.

    Then, compute the upper- and lower-limit for angle Ⓒ and dimensions Ⓐ, Ⓓ, Ⓔ, Ⓕ, and Ⓖ.

1. _____

2. _____

3. _____

4. _____

5. _____

6. Ⓒ = _____

7. _____

8. _____

9. _____

10. _____

11. Ⓐ = _____
    Ⓓ = _____

12. _____

13. _____

14. _____

15. Ⓔ = _____

16. Ⓕ = _____

17. Ⓖ = _____

18. _____

19. AL [ ] STL [ ]

20. (a) _____
    (b) _____

21. Ⓒ = _____
    Ⓐ = _____
    Ⓓ = _____
    Ⓔ = _____
    Ⓕ = _____
    Ⓖ = _____

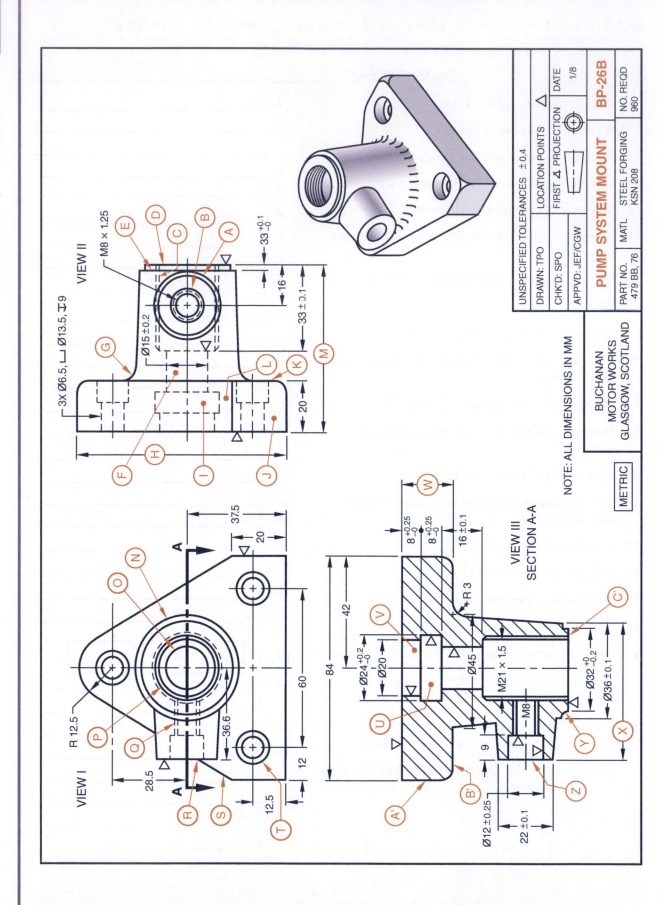

VIEW II

M8 × 1.25

3X Ø6.5, ⌴ Ø13.5, ⊼ 9

Ø15 ±0.2

33 +0.1/−0

16

33 ± 0.1

20

M

A B C D E G H F I K L J

VIEW I

37.5

20

28.5

60

36.6

12

12.5

R 12.5

A

A

N O P Q R S T

SECTION A-A
VIEW III

8 +0.25/−0

8 +0.25/−0

16 ± 0.1

R 3

42

84

9

22 ± 0.1

Ø24 +0.2/−0

Ø20

Ø45

M21 × 1.5

M8

Ø32 +0/−0.2

Ø36 ± 0.1

Ø12 ± 0.25

W V U A' B' X Y Z C'

NOTE: ALL DIMENSIONS IN MM

METRIC

UNSPECIFIED TOLERANCES ±0.4		
DRAWN: TPO	LOCATION POINTS	
CHK'D: SPO	FIRST ∡ PROJECTION	DATE
APPVD: JEF/CGW	⊕	1/8

PUMP SYSTEM MOUNT

BP-26B

| PART NO. 479 BB, 76 | MATL STEEL FORGING KSN 208 | NO. REQD 960 |

BUCHANAN
MOTOR WORKS
GLASGOW, SCOTLAND

ASSIGNMENT—UNIT 26B: PUMP SYSTEM MOUNT (BP-26B)

Student's Name _____

1. State what system of projection is used.

2. Name VIEWS I, II, and III.

3. Give the symbol that indicates a diameter.

4. Tell what the symbol △ indicates.

5. Indicate what tolerance to use if none is given.

6. Give the specifications for the counterbored holes Ⓣ.

7. Determine the upper- and lower-limit dimensions for diameters Ⓐ and Ⓑ.

8. Give the ISO designation for threads Ⓒ. Describe what each value means.

9. Locate surfaces Ⓓ and Ⓔ in VIEW III.

10. State the lower-limit diameter for hole Ⓕ.

11. Give the radius of fillet Ⓖ.

12. Compute maximum overall dimension Ⓗ.

13. Give the minimum width of undercut Ⓘ.

14. Locate surfaces Ⓙ and Ⓚ in VIEW III and surface Ⓛ in VIEW I.

15. Compute maximum overall dimension Ⓜ, adding all the + tolerances to the basic dimensions.

16. Give minimum diameters Ⓝ and Ⓞ.

17. State the thread size for Ⓠ.

18. Locate surface Ⓡ in VIEW III.

19. Give the lower-limit diameter and width of holes Ⓤ and Ⓥ.

20. Determine upper-limit overall dimensions of Ⓦ and Ⓧ.

21. Determine the basic dimension between surfaces Ⓔ and Ⓚ.

1. _____

2. I = _____ II = _____ III = _____

3. _____

4. _____

5. _____

6. _____

7. Upper Ⓐ = _____ Ⓑ = _____

   Lower Ⓐ = _____ Ⓑ = _____

8. _____

9. Ⓓ = _____

10. Ⓕ = _____

11. Ⓖ = _____

12. Ⓗ = _____

13. Ⓘ = _____

14. Ⓙ = _____ Ⓚ = _____

15. Ⓜ = _____

16. Ⓝ = _____ Ⓞ = _____

17. Ⓠ = _____

18. Ⓡ = _____

19. Ⓤ = _____ Ⓥ = _____

20. Ⓦ = _____ Ⓧ = _____

21. Basic dimension = _____

Half Sections, Partial Sections, and Full Section Assembly Drawings

HALF SECTIONS

The internal and external details of a part may be represented clearly by a sectional view called a half section. In a half-section view, one-half of the object is drawn in section and the other half is drawn as an exterior view, Figure 27–1. The half section is used principally where both the inside and outside details are symmetrical and where a full section would omit some important details in an exterior view.

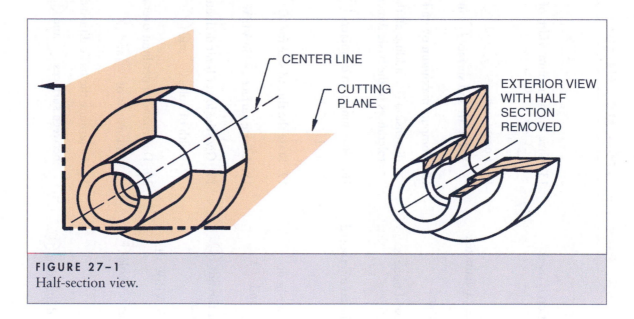

CENTER LINE

CUTTING PLANE

EXTERIOR VIEW WITH HALF SECTION REMOVED

FIGURE 27–1
Half-section view.

Theoretically, the cutting plane for a half-section view extends halfway through the object, stopping at the axis or center line, Figure 27–2. Drawings of simple symmetrical parts may not always include the cutting plane line or the arrows and letters showing the direction in which the section is taken. Also, hidden lines are not shown in the sectional view unless they are needed to give details of construction or for dimensioning.

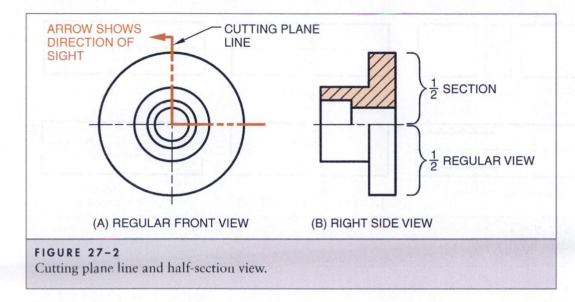

FIGURE 27-2
Cutting plane line and half-section view.

BROKEN OR PARTIAL SECTIONS

On some parts, it is not necessary to use either a full or a half section to expose the interior details. In such cases, a broken-out or partial section may be used, Figure 27–3. The cutting plane is imagined as being passed through a portion of the object, and the part in front of the plane is then broken away. The break line is an irregular freehand line that separates the internal sectioned view and the external view.

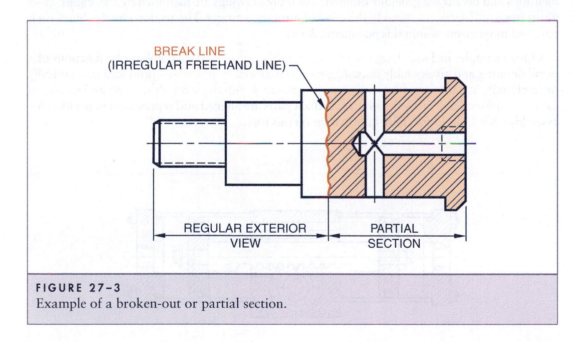

FIGURE 27-3
Example of a broken-out or partial section.

CONVENTIONAL BREAKS

A long part with a uniform cross section may be drawn to fit on a standard size drawing sheet by cutting out a portion of the length. In this manner, a part may be drawn larger to bring out some complicated details.

The cutaway portion may be represented by a conventional symbol that does two things: (1) it indicates that a portion of the uniform cross section is removed, and (2) it shows the internal shape. A few conventional symbols that are accepted as standard are illustrated in Figure 27–4.

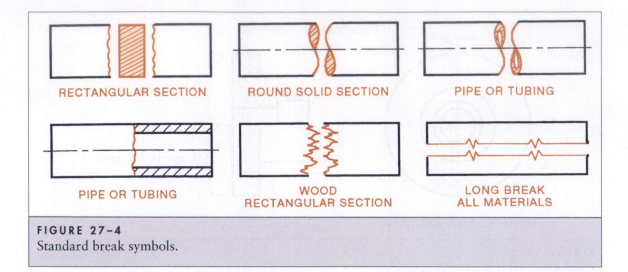

FIGURE 27-4
Standard break symbols.

ASSEMBLY DRAWINGS: TYPES AND FUNCTIONS

Full, half, partial, and cutaway sections are often included in assembly drawings to display the external shape of a portion of a complete mechanism as well as internal features and movements of parts that make up the mechanism. Assembly drawings are valuable in design, construction, assembly, and preparation of specifications.

An assembly drawing may include one or more standard views, with or without sections. Dimensions and details are generally omitted. Assembly drawings are identified by use. Figure 27–5 illustrates a full section assembly drawing of a rotary actuator. The section clearly shows each part and movements within this pneumatic device.

Other examples include: design working assembly drawings (that combine the functions of a detail drawing and an assembly drawing); pictorial assembly drawings (produced mechanically or freehand); installation and maintenance assembly; subassembly drawings; and technical catalog exploded assembly drawings (in which parts are aligned and represented in position for assembly). See Unit 33 for more information on this topic.

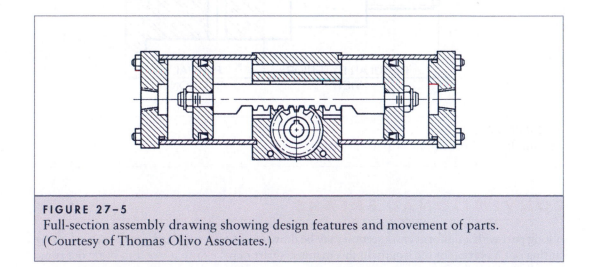

FIGURE 27-5
Full-section assembly drawing showing design features and movement of parts.
(Courtesy of Thomas Olivo Associates.)

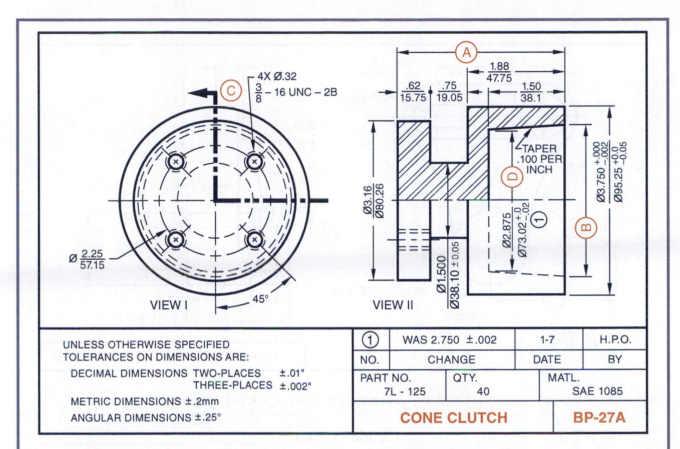

VIEW I

VIEW II

①	WAS 2.750 ±.002	1-7	H.P.O.
NO.	CHANGE	DATE	BY

PART NO.	QTY.	MATL.
7L - 125	40	SAE 1085

UNLESS OTHERWISE SPECIFIED TOLERANCES ON DIMENSIONS ARE:

DECIMAL DIMENSIONS TWO-PLACES   ±.01"
                                    THREE-PLACES ±.002"

METRIC DIMENSIONS ±.2mm

ANGULAR DIMENSIONS ±.25°

CONE CLUTCH   **BP-27A**

ASSIGNMENT A—UNIT 27: CONE CLUTCH (BP-27A)

Student's Name _____

NOTE: Compute all dimensions, where applicable, in customary inch and SI metric (mm) values.

1. Name VIEW I and VIEW II in third-angle projection.

2. What type of line is Ⓒ?

3. Give the threaded holes specification.

4. What type of screw thread representation is used?

5. Give the inch diameter on which the threaded holes are located.

6. Give the upper-limit dimension for the 45° angle.

7. Determine basic overall length Ⓐ.

8. Give the lower-limit for the 1.500" DIA.

9. Compute basic dimension Ⓑ.

10. Give the original diameter of Ⓓ.

11. Change the tolerances as follows: Two-place decimal dimensions $^{+.00}_{-.02}$; three-places, $^{+.000}_{-.0025}$"; millimeter dimensions $^{+0.0}_{-0.5}$; angular dimension $^{+.08°}_{-.25}$. Then, compute the upper- and lower-limits for the 45° angle and dimensions Ⓐ, Ⓑ, and Ⓓ.

1. (I) _____ (II) _____

2. _____

3. _____

4. _____

5. _____

6. _____

7. Ⓐ = _____

8. _____

9. Ⓑ = _____

10. Ⓓ = _____

11. 45° = _____ Min. _____ Max.

	Minimum		Maximum	
	inch	mm	inch	mm
Ⓐ =				
Ⓑ =				
Ⓓ =				

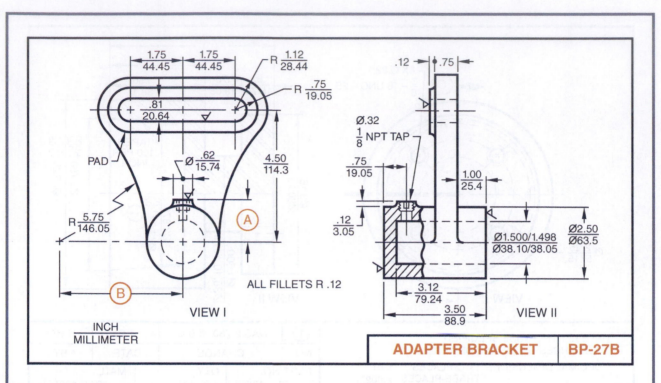

1.75 / 44.45   1.75 / 44.45   R 1.12 / 28.44

R .75 / 19.05

.81 / 20.64

PAD

Ø .62 / 15.74   4.50 / 114.3

R 5.75 / 146.05

A

B

INCH / MILLIMETER

VIEW I

ALL FILLETS R .12

.12   .75

Ø.32
1/8 NPT TAP

.75 / 19.05   1.00 / 25.4

.12 / 3.05

Ø1.500/1.498 / Ø38.10/38.05   Ø2.50 / Ø63.5

3.12 / 79.24

3.50 / 88.9

VIEW II

ADAPTER BRACKET   **BP-27B**

ASSIGNMENT B—UNIT 27: ADAPTER BRACKET (BP-27B)

Student's Name _____

NOTE: Compute all dimensions, where applicable, in customary inch and SI metric (mm) values.

1. Name VIEW I and VIEW II (third-angle projection).

2. How many outside machined surfaces are indicated in **VIEW II?**

3. Give the upper- and lower-limits of the bored hole.

4. How deep is the hole bored?

5. What is the drill size for the tapped hole?

6. What size pipe tap is used?

7. Give the basic center-to-center distance of the elongated slot.

8. Compute basic dimensions (A).

9. Determine the overall basic length of the pad.

10. Compute basic dimension (B).

11. Add a $^{+.01''}_{-.00''}$ tolerance to all two-place decimal dimensions and $^{+0.2}_{-0.0}$ mm to all metric dimensions. Then, determine the upper- and lower-limit dimensions for:

   (a) The depth of the bored hole.

   (b) The center-to-center distance between the elongated slot and the bored hole.

   (c) The overall width of the bracket (VIEW I).

   (d) The overall bracket height.

   (e) Dimensions (A) and (B).

1. (I) _____ (II) _____

2. _____

3. Upper _____ " _____ mm

   Lower _____ " _____ mm

4. _____ " _____ mm

5. _____

6. _____

7. _____ " _____ mm

8. (A) = _____ " _____ mm

9. _____

10. (B) = _____

	Minimum		Maximum	
	inch	mm	inch	mm
11. (a)				
(b)				
(c)				
(d)				
(e) (A)				
(B)				

Computer Numerical Control (CNC) Fundamentals

Datums: Ordinate and Tabular Dimensioning

COMPUTER NUMERICAL CONTROL

Machines controlled by data that are specially coded with numbers, letters, and special characters are operated by numerical control (NC). This encoded information defines the path of the machine tool in accomplishing specific tasks. These instructions can be fed into the machine from several different sources—an operator, a computer, a computer network, or portable magnetic media. When the job changes, the instructions must be rewritten to accommodate these changes. This is a relatively easy job and this flexibility is the key to the growth of numerical control.

Actually, NC is a broad term that includes traditional approaches as well as more modern developments, such as computer numerical control (CNC) and direct numerical control (DNC).

CNC processing and distribution applications are numerous, covering every material imaginable: metals, plastics/polymers, ceramics, composites, stone, wood, fabrics, leather, etc. See Units 30 and 45 for additional CNC information.

DATUMS

A datum is a reference point, axis, or plane from which dimensions are located and essential information is derived. A datum may be a point, line, plane, cylinder, or other exact feature. Datums are used in design, manufacturing, CNC, and many other processes. Datums have indispensable roles to perform in geometric dimensioning and tolerancing, as will be learned in Unit 29.

When a datum is specified, such as the two that are illustrated in Figure 28–1, all features must be stated in relation to the datum reference. Different features of the part are located from the datum, not from other features. Datums must be clearly identified or easily recognizable. At least two, and often three datums must be used to define or measure a part. Because all measurements are taken from datums (similar to those earlier used in base line dimensioning), errors in dimensioning are not cumulative.

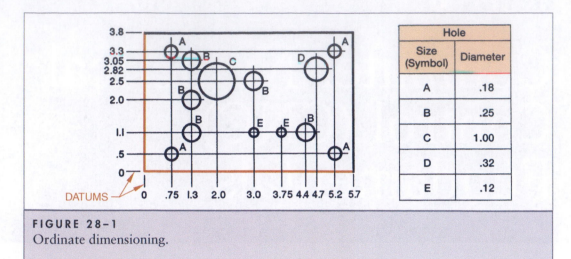

Hole	
Size (Symbol)	Diameter
A	.18
B	.25
C	1.00
D	.32
E	.12

FIGURE 28–1
Ordinate dimensioning.

ORDINATE DIMENSIONING AND NUMERICAL CONTROL MACHINE DRAWINGS

Ordinate dimensioning is a type of rectangular datum dimensioning. In ordinate dimensioning, the dimensions are measured from two or three mutually related datum planes, Figure 28–1. The datum planes are indicated as zero coordinates. Dimensions from the zero coordinates are represented on drawings as extension lines, without the use of dimension lines or arrowheads.

Specific features are located by the intersection of datum dimensions. Ordinate dimensioning is used when there would otherwise be a large number of dimensions and features and close tolerances are required. Datums and ordinate dimensioning are the foundation of numerically controlled machine drawings and precision parts.

TABULAR DIMENSIONING

Tabular dimensioning is another form of rectangular datum dimensioning. As the term implies, dimensions from intersecting datum planes are given in a table, Figure 28–2. Tabular dimensioning helps to eliminate possible errors that could result from incorrectly reading a dimension when a large number is included on a drawing.

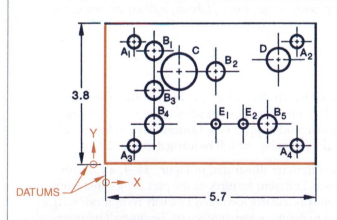
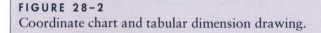

Hole		Location (Axes)	
Identification	Size	X →	Y ↑
A₁	.18	.75	3.3
A₂	.18	5.2	3.3
A₃	.18	.75	.5
A₄	.18	5.2	.5
B₁	.25	1.3	3.05
B₂	.25	3.0	2.5
B₃	.25	1.3	2.0
B₄	.25	1.3	1.1
B₅	.25	4.4	1.1
C	1.00	2.0	2.5
D	.32	4.7	2.82
E₁	.12	3.0	1.1
E₂	.12	3.75	1.1

FIGURE 28–2
Coordinate chart and tabular dimension drawing.

Values of **X** and **Y** are measured along the respective datum lines from their intersection. The intersection is the origin of the datums. Tabular dimensioning is recommended when there are a great many repetitive features. These would make a dimensioned drawing difficult to read. Dimensions may also be given on a second or additional view for operations to be performed along **Z** or other axes.

Datums have indispensable roles to perform in geometric dimensioning and tolerancing, as will be learned in Unit 29.

COORDINATE CHARTS

The dimensions of a coordinate and features that are to be fabricated on coordinates are given in a coordinate chart, Figure 28–2. Only those dimensions that originate at the datums are found in such a chart. The coordinates may specify sufficient information to fabricate the entire feature or only a portion of it.

Parts or components may be represented on drawings by ordinate or tabular dimensioning in the English or metric system of measurement or in both systems.

CNC CHAIN DIMENSIONING

There are two reference point systems used to position tools in a CNC machine for work on a part.

In absolute positioning, all process locations are given as distances from a zero point (datum). Each move the tool makes is given as a distance and direction from the zero point as would be the case in tabular and ordinate dimensioning.

The second method is called incremental positioning, which uses successive dimensions or chain dimensions, Figure 28–3. Each move of the tool moves a specific distance and direction from its present location to the next location without returning to zero.

Individual CNC machine tolerances are accounted for in the operational program. The machine tool moves, following basic dimensions; therefore, tolerances would not build up. Theoretically, each part produced would be the exact same size and create fewer rejected interchangeable parts.

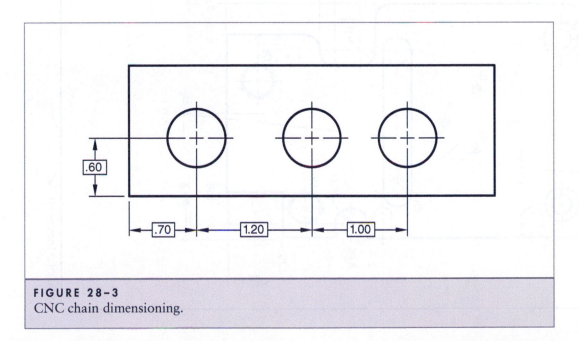

FIGURE 28–3
CNC chain dimensioning.

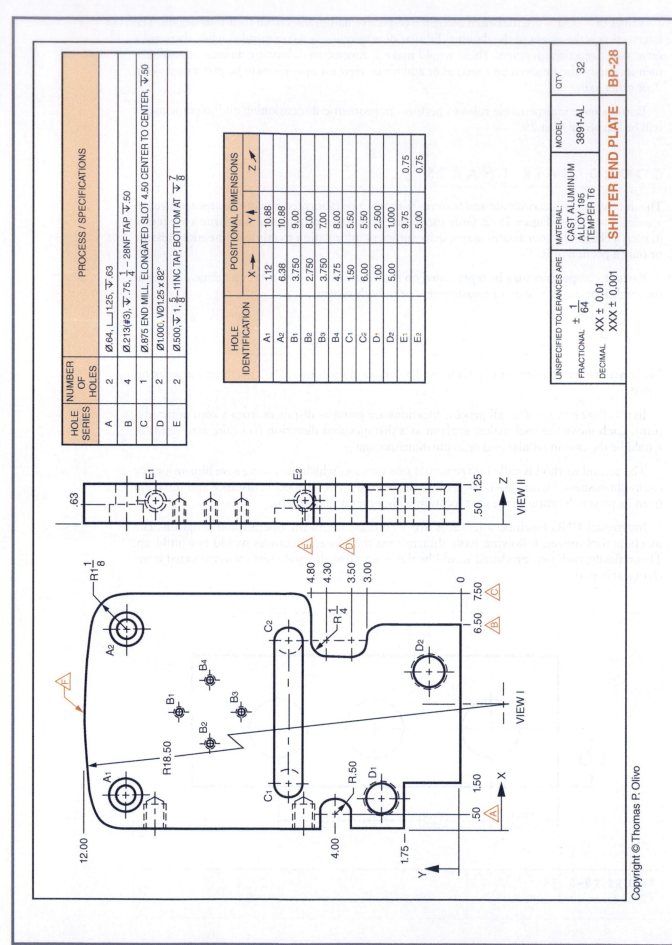

HOLE SERIES	NUMBER OF HOLES	PROCESS / SPECIFICATIONS
A	2	Ø.64, ⌴1.25, ⍌.63
B	4	Ø.213(#3), ⍌.75, $\frac{1}{4}$ – 28NF TAP ⍌.50
C	1	Ø.875 END MILL, ELONGATED SLOT 4.50 CENTER TO CENTER, ⍌.50
D	2	Ø1.000, VØ1.25 x 82°
E	2	Ø.500 ⍌ 1, $\frac{5}{8}$–11NC TAP, BOTTOM AT ⍌ $\frac{7}{8}$

HOLE IDENTIFICATION	POSITIONAL DIMENSIONS X →	Y ↑	Z ↗
A₁	1.12	10.88	
A₂	6.38	10.88	
B₁	3.750	9.00	
B₂	2.750	8.00	
B₃	3.750	7.00	
B₄	4.75	8.00	
C₁	1.50	5.50	
C₂	6.00	5.50	
D₁	1.00	2.500	
D₂	5.00	1.000	
E₁		9.75	0.75
E₂		5.00	0.75

UNSPECIFIED TOLERANCES ARE	MATERIAL: CAST ALUMINUM ALLOY 195 TEMPER T6	MODEL 3891-AL	QTY 32
FRACTIONAL ± $\frac{1}{64}$ DECIMAL XX ± 0.01 XXX ± 0.001	SHIFTER END PLATE		BP-28

ASSIGNMENT—UNIT 28: SHIFTER END PLATE (BP-28)

Student's Name _____

1. (a) Give the material specifications for the shifter end plate.

   (b) Indicate the model number and required quantity.

2. (a) Name VIEW I and VIEW II.

   (b) State briefly why VIEW II is used.

3. Give the Basic overall sizes for the width, height, and depth of the plate.

4. Classify dimensions (A), (B), (C), (D), and (E) according to type.

5. (a) Identify the two dimensioning methods used on the shifter end plate drawing.

   (b) State two distinguishing characteristics for arrowless dimensioning.

6. (a) Give the number of datums used.

   (b) Indicate how each datum is specified.

7. (a) State the number of holes that are to be machined for each series of holes.

   (b) Name the hole-forming process(es) required to machine each hole series.

8. Determine the center (position) dimensions for holes D_1 and D_2.

9. (a) Give the locational dimensions for the center of C_1 to start the end-milling cut for the elongated slot.

   (b) Determine (1) the basic height and (2) basic overall width of the slot.

   (c) State the minimum and the maximum dimensions for the height and width of the slot.

10. Give the maximum radius for (a) arc (F) and (b) the rounded corners on the arc portion.

11. Compute the minimum center-to-center Y axis dimensions between the following holes:

    (a) B_1 and B_2 (c) D_1 and D_2
    (b) B_1 and B_3 (d) E_1 and E_2

12. Determine the following maximum and minimum depths along the Z axis:

    (a) The elongated slot   (b) The counterbored hole.

---

1. (a) _____

   (b) Model _____ Quantity _____

2. (a) VIEW I _____
       VIEW II _____

   (b) _____
       _____

3. Maximum: width _____ ;
   height _____ ; depth _____

4.

Dimension	Size	Type	
			Position
(A)			
(B)			
(C)			
(D)			
(E)			

5. (a) 1 _____
       2 _____

   (b) 1 _____
       2 _____

6. (a) _____

   (b) _____

7.

Series	# Holes	Matching Process(es)

8.   X Axis   Y Axis

   D_1 _____ _____

   D_2 _____ _____

9. (a) C_1 _____ C_2 _____

   (b) Basic Height = _____

       Basic Width = _____

              Height   Width

   (c) _____ _____

       Minimum _____

       Maximum _____

10. (a) Maximum _____

    (b) Maximum Corners _____

11. Y Axis Minimum Dimensions

    (a) _____   (c) _____

    (b) _____   (d) _____

12.   Max. Depth   Min. Depth

    (a) Slot _____ _____

    (a) C'bored _____ _____

Geometric Dimensioning and Tolerancing

29

Geometric Dimensioning, Tolerancing, and Datum Referencing

Geometric dimensioning and tolerancing (GD&T) is a three-dimensional mathematical system. It is an exact method of dimensioning using datums. GD&T provides the largest amount of permissible variation in form (shape and size) and position (location), allowing worldwide interchangeability of parts. With the increasing use of Computer Aided Manufacturing, as discussed in Unit 30, GD&T will be used more and more; however, do not expect to see GD&T on every blueprint. It's only necessary when geometric accuracy must be greater than ordinarily expected from the manufacturing processes.

Tolerances of form relate to straightness, flatness, roundness, cylindricity, perpendicularity, angularity, and parallelism. Tolerances of position relate to the location of such features as holes, grooves, stepped areas, and other details of a part.

Dimensioning and tolerancing, as covered in earlier units, relate to acceptable limits of deviation represented on drawings by using unilateral, bilateral, and upper- and lower-limit tolerancing. By contrast, geometric dimensioning and tolerancing provide for all characteristics of a part in terms of function, relationship to other mating parts, and to design, production, measurement, inspection, assembly, and operation.

TYPES, CHARACTERISTICS, AND SYMBOLS

The five basic types of tolerances are: (1) **form**, (2) **profile**, (3) **orientation**, (4) **location**, and (5) **runout.** Some tolerances relate to individual or related features of a part; others to a combination of individual and related features (Figure 29–1A). DATUM requirements and other characteristics are identified by modifying symbols (Figure 29–1B) that are added following a tolerance value to indicate how the tolerance is modified. There are five basic modifying symbols: Ⓜ, Ⓛ, Ⓕ, Ⓣ, and Ⓟ. However, for purposes of this text, only modifier Ⓜ will be used. Maximum material condition (MMC) implies that a tolerance, modified by the symbol Ⓜ, requires a part to be machined to a size limit where it contains the maximum amount of material. In other words, the MMC of a shaft of .750″ diameter with $^{+.002}_{-.002}$ tolerancing contains the maximum amount of material when it measures .752″. The least material condition (LMC), identified by the symbol Ⓛ, would be .748″.

(A) Geometric Tolerancing Characteristics and Symbols

Datum Requirement	Features	Type of Tolerance	Characteristic	Symbol
Never Needed	Individual Features	Form	Straightness	—
			Flatness	▱
			Roundness (Circularity)	○
			Cylindricity	⌭
Always Needed	Related Features	Orientation	Angularity	∠
			Perpendicularity	⊥
			Parallelism	//
		Location	Position	⊕
			Concentricity	◎
			Symmetry	≡
		Runout	Circular Runout	↗ ↗
			Total Runout	⌰ ⌰
Required Most of the Time	Individual or Related Features	Profile	Profile of a Line	⌒
			Profile of a Surface	⌓

(B) Selected Modifying Terms and Symbols

Term (abbreviation)	Symbol
Maximum Material Condition (MMC)	Ⓜ
Least Material Condition (LMC)	Ⓛ
Free State	Ⓕ
Projected Tolerance Zone	Ⓟ
Tangent Zone	Ⓣ

FIGURE 29–1
Geometric tolerancing characteristics, modifying terms, symbols, and datum requirements.

DATUM REFERENCE FRAME AND PART FEATURES

To review, a datum identifies the origin of a dimensional relationship between a particular (designated) point or surface and a measurement. In geometric dimensioning and tolerancing, datums are identified in relation to a theoretical reference frame. The datum reference frame (similar to a drawing projection box) consists of three mutually perpendicular planes: first datum plane, second datum plane, and third (tertiary) datum plane, Figure 29–2. The letters A, B, and C indicate the sequence (order of importance) of the datums; the arrowheads, the direction of the measurements.

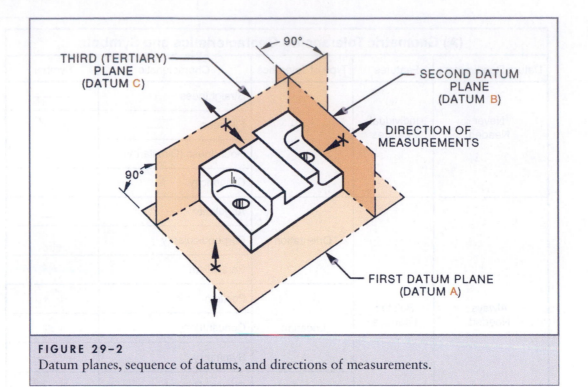

FIGURE 29–2
Datum planes, sequence of datums, and directions of measurements.

DATUM FEATURES OF CYLINDRICAL PARTS

Datums for cylindrical parts are generally related to three datum planes. Datum A identifies the base of the part. Datum B refers to the axis or center line. These datums are shown in Figure 29–3. When used, Datum C is perpendicular to Datum A and parallel to Datum B.

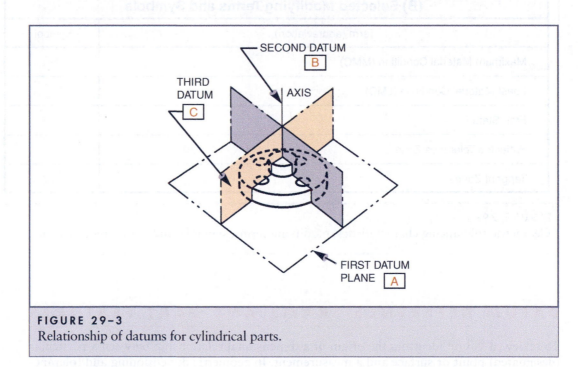

FIGURE 29–3
Relationship of datums for cylindrical parts.

DATUM FEATURE AND TARGET SYMBOLS

A datum feature symbol consists of a filled or unfilled triangular base and a lead line attached to a datum labeled rectangle (see Figure 29–4A). ISO datum featured symbols have a 90° triangular base, while ASME have a 60° base.

A datum target symbol consists of a circle divided horizontally to form two half circles (Figure 29–4B). A radial leader is used to reference the datum target symbol and contents to a specific surface, point, or line on a drawing. The size and location of the target area may be included using basic dimensioning practices.

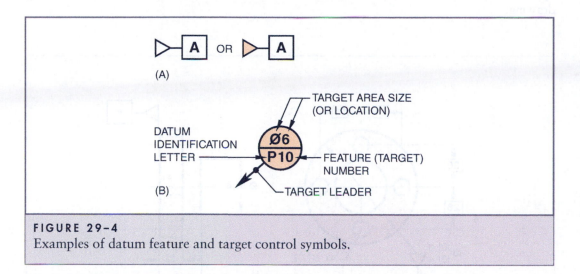

FIGURE 29–4
Examples of datum feature and target control symbols.

FEATURE CONTROL FRAME

A feature control frame provides a convenient, accurate method for incorporating geometric characteristic symbols, tolerance values, modifying symbols, datum, and dimensioning and tolerancing information. The control frame may contain one or more compartments in which control information is identified following a specified sequence (Figure 29–5).

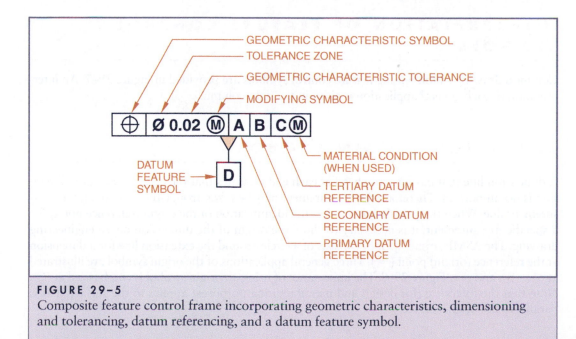

FIGURE 29–5
Composite feature control frame incorporating geometric characteristics, dimensioning and tolerancing, datum referencing, and a datum feature symbol.

The first entry in a feature control frame is the symbol of the geometric characteristic. The second entry provides the tolerancing dimension. When applicable, the symbol for material condition is the third entry. The tolerancing and material condition symbols are contained in one compartment of the frame. This information is followed by one or more additional compartments, depending on the number of datum reference letters (symbols) that are used. In instances when a datum is established from two datum features, both reference letters are separated by a dash and appear in a compartments, for example, // .05 A-B

Figure 29–5 provides a summary of how datums as well as geometric dimensioning and tolerancing symbols and concepts are contained in a feature control frame. Figure 29–6 illustrates the application of datum feature symbols A and B and a feature control frame on an engineering drawing.

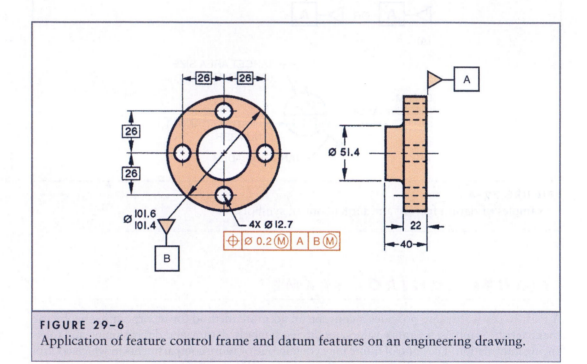

FIGURE 29–6
Application of feature control frame and datum features on an engineering drawing.

INTERPRETATION OF FEATURE CONTROL SYMBOLS

Common drawing callouts using feature control symbols are provided in Figure 29–7. An interpretation of each general application is given in the right column.

DIMENSION ORIGIN SYMBOL

A dimension line that has an arrowhead at each end indicates that the origin point of a dimension is not important. Therefore, the measurement may be taken from either end along the measurement line. When the part design requires the identification of the origin (reference point) for a specific measurement, it is necessary to show the origin of the dimension on an engineering drawing. The ASME origin symbol consists of a circle around the extension line for a dimension at the reference (origin) point (⊕). Three general applications of the origin symbol are illustrated at (A), (B), and (C), Figure 29–8. The importance of using an origin symbol is emphasized at (C). Note that the tolerance (for design and manufacturing purposes) applies to the surface that is parallel to the origin surface.

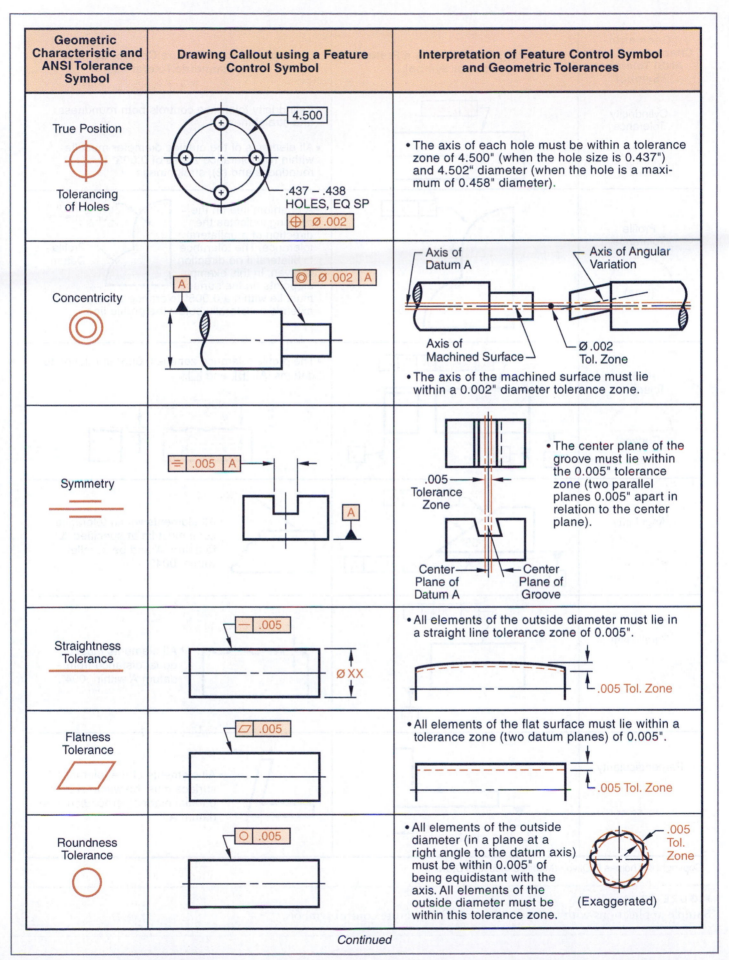

Geometric Characteristic and ANSI Tolerance Symbol	Drawing Callout using a Feature Control Symbol	Interpretation of Feature Control Symbol and Geometric Tolerances
True Position ⊕ Tolerancing of Holes	4.500 .437 – .438 HOLES, EQ SP ⊕ Ø .002	• The axis of each hole must be within a tolerance zone of 4.500" (when the hole size is 0.437") and 4.502" diameter (when the hole is a maximum of 0.458" diameter).
Concentricity ◎	A ◎ Ø .002 A	Axis of Datum A — Axis of Angular Variation — Axis of Machined Surface — Ø .002 Tol. Zone • The axis of the machined surface must lie within a 0.002" diameter tolerance zone.
Symmetry	≡ .005 A A	.005 Tolerance Zone — Center Plane of Datum A — Center Plane of Groove • The center plane of the groove must lie within the 0.005" tolerance zone (two parallel planes 0.005" apart in relation to the center plane).
Straightness Tolerance	— .005 Ø XX	• All elements of the outside diameter must lie in a straight line tolerance zone of 0.005". .005 Tol. Zone
Flatness Tolerance	▱ .005	• All elements of the flat surface must lie within a tolerance zone (two datum planes) of 0.005". .005 Tol. Zone
Roundness Tolerance ○	○ .005	• All elements of the outside diameter (in a plane at a right angle to the datum axis) must be within 0.005" of being equidistant with the axis. All elements of the outside diameter must be within this tolerance zone. .005 Tol. Zone (Exaggerated)

Continued

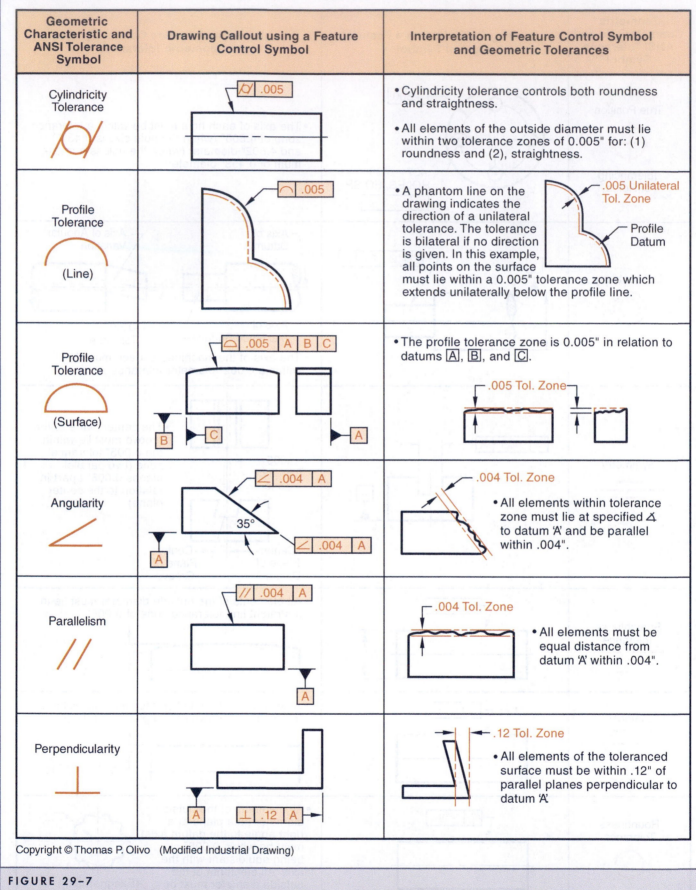

Geometric Characteristic and ANSI Tolerance Symbol	Drawing Callout using a Feature Control Symbol	Interpretation of Feature Control Symbol and Geometric Tolerances
Cylindricity Tolerance	⌀ .005	• Cylindricity tolerance controls both roundness and straightness. • All elements of the outside diameter must lie within two tolerance zones of 0.005" for: (1) roundness and (2), straightness.
Profile Tolerance (Line)	⌒ .005	• A phantom line on the drawing indicates the direction of a unilateral tolerance. The tolerance is bilateral if no direction is given. In this example, all points on the surface must lie within a 0.005" tolerance zone which extends unilaterally below the profile line. .005 Unilateral Tol. Zone Profile Datum
Profile Tolerance (Surface)	⌒ .005 A B C	• The profile tolerance zone is 0.005" in relation to datums Ⓐ, Ⓑ, and Ⓒ. .005 Tol. Zone
Angularity	∠ .004 A 35° ∠ .004 A	.004 Tol. Zone • All elements within tolerance zone must lie at specified ∠ to datum 'A' and be parallel within .004".
Parallelism	∥ .004 A	.004 Tol. Zone • All elements must be equal distance from datum 'A' within .004".
Perpendicularity	⊥ .12 A	.12 Tol. Zone • All elements of the toleranced surface must be within .12" of parallel planes perpendicular to datum 'A'

Copyright © Thomas P. Olivo (Modified Industrial Drawing)

FIGURE 29-7
Sample applications with interpretations of ANSI feature control symbols.

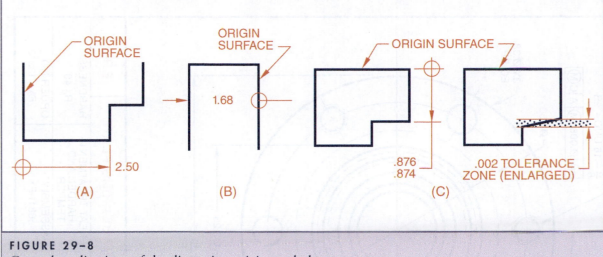

FIGURE 29-8
General applications of the dimension origin symbol.

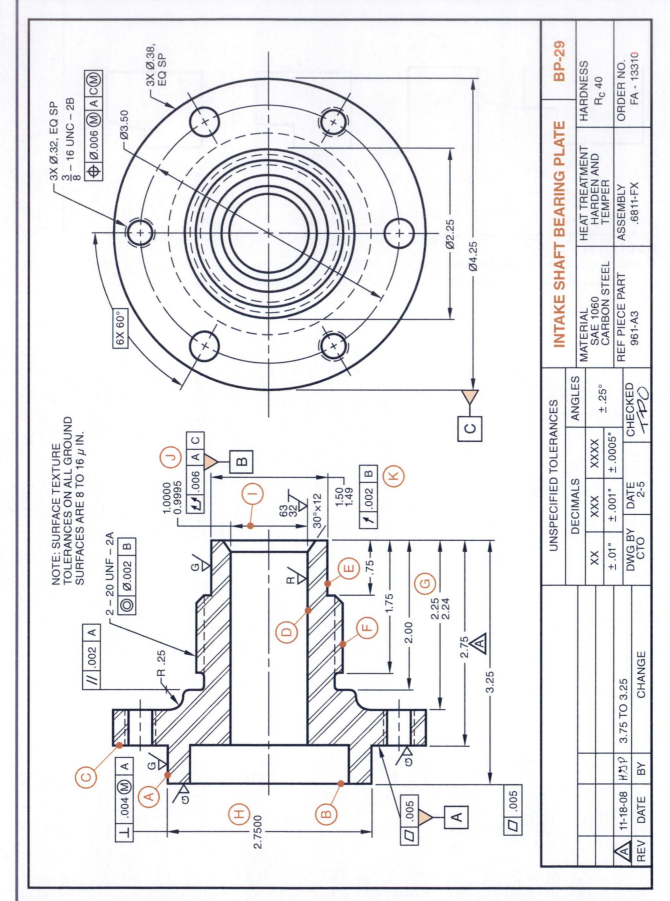

NOTE: SURFACE TEXTURE
TOLERANCES ON ALL GROUND
SURFACES ARE 8 TO 16 μ IN.

3X Ø.32, EQ SP
3/8 – 16 UNC – 2B

3X Ø.38,
EQ SP

2 – 20 UNF – 2A

6X 60°

Ø3.50

Ø2.25

Ø4.25

	INTAKE SHAFT BEARING PLATE		BP-29
	MATERIAL SAE 1060 CARBON STEEL	HEAT TREATMENT HARDEN AND TEMPER	HARDNESS R$_C$ 40
	REF PIECE PART 961-A3	ASSEMBLY .6811-FX	ORDER NO. FA - 13310

UNSPECIFIED TOLERANCES			ANGLES
DECIMALS			± .25°
XX	XXX	XXXX	
± .01"	± .001"	± .0005"	
DWG BY CTO	DATE 2-5	CHECKED TPO	

/△\	11-18-08	HJP	3.75 TO 3.25
REV	DATE	BY	CHANGE

2.7500

ASSIGNMENT—UNIT 29: INTAKE SHAFT BEARING PLATE (BP-29)

Student's Name _____

1. Determine the basic (a) outside diameter and (b) width of the intake shaft bearing plate.

2. (a) List the basic thread specifications for Ⓕ and (b) give the meaning of each value or symbol in the specifications.

3. Give the specifications for the three threaded (tapped) holes.

4. List the tolerance on unspecified (a) XX, XXX, and XXXX place decimal dimensions and (b) angular dimensions.

5. Identify the machining processes for (a) surfaces and Ⓐ, Ⓑ, Ⓒ and, Ⓔ (b) feature Ⓓ.

6. Give the surface texture tolerances for the ground and reamed features.

7. Determine the upper- and lower-limit diameters of (a) Ⓗ and (b) reamed hole Ⓘ.

8. Give one main advantage of geometric dimensioning and tolerancing over earlier conventional standards.

9. (a) List four basic types of geometric tolerancing found on the drawing (BP-29).

   (b) Identify the characteristic represented on the drawing by each type of tolerancing.

   (c) Draw the symbol for each identified characteristic.

10. Show how datum feature symbol Ⓒ is represented on the drawing.

11. Draw a datum target symbol that contains the following geometric tolerancing information:

    (a) Datum identification F

    (b) Feature target #3

    (c) Target Area 6

12. Interpret the meaning of each symbol or dimension within the compartments of the following control frame: ⊕ ⟋ .005

13. (a) Draw the feature control frame, including all geometric tolerancing information, for thread feature Ⓕ.

    (b) Give the meaning of each symbol and dimensional value for each compartment of the frame. Ⓙ & Ⓚ.

14. Interpret the meaning of the feature control frames Ⓙ & Ⓚ.

15. Briefly state what a dimension origin symbol (⊕) on a blueprint indicates in relation to a tolerance.

	(a) Type	(b) Characteristic	(c) Symbol
9.			
(1)	_____	_____	_____
(2)	_____	_____	_____
(3)	_____	_____	_____
(4)	_____	_____	_____

10. _____

11. _____

12. _____

13. (a)		
	Symbol	Meaning
(b)	_____	_____

14. Ⓙ _____

15. _____

Ⓚ _____

1. (a) Basic outside diameter _____
   (b) Basic width _____

2. (a) _____ (b) _____

3. _____

4. (a) XX _____ XXX _____ XXXX _____
   (b) _____

5. (a) _____ (b) _____

6. _____

7. | | Upper | Lower |
   |---|---|---|
   | (a) Ⓗ | | |
   | (b) Ⓘ | | |

8. _____

Computer Graphics Technology

30

CADD/CAM/CIM and ROBOTICS

Computer graphics broadly identifies drawings, three-dimensional displays, color animation, and other visual products drawn with the aid of a computer. **Computer-Aided Drafting (CAD)** is a computer graphics system. **CADD (Computer-Aided Drafting and Design)** is another computer graphics system. Using this type of computer graphics software, drawings in 3rd ∡ projection can easily be changed to 1st ∡ projection and vice-versa.

CAPABILITIES OF CAD SYSTEMS

CAD systems utilize all the principles and applications of drafting techniques, engineering data, and information generally found on drawings produced by conventional methods. CAD can make every type of drawing that can be done manually.

CAD scaled drawings may be enlarged or reduced and automatically dimensioned to provide necessary manufacturing information. Cross-hatching for sectional views is produced by selecting the correct cross-hatch lines for the part material, choosing the desired angle for the lines, and defining the area to be cross-hatched. Single, multiple, and/or auxiliary views may be drawn using CAD. A completed drawing may be reproduced in black and white or color, stored in memory, recalled at any time, and edited for changes, corrections, or additions.

Details are generated by **zoning in** on a feature and **zoning out** to view the drawing in full. Notes and straight, slant, and architectural style letters and numbers may be selected and placed at desired locations on drawings.

CAD has several claimed advantages over manual drafting. CAD is said to be more accurate, faster in many instances, neater, and more consistent in quality. Some of the most important advantages are in the areas of immediate retrieval, revision, and storage of drawings.

CADD AND CAM

The functions generated by CADD are handled through a **central processing unit (CPU)** or computer. The computer includes a number of **integrated-circuit chips,** each of which controls a specific function. CADD, therefore, has added capability to perform complex mathematical processes, generate engineering data, and process commands to control tooling, manufacturing, architectural planning, electronic circuitry, and other automated production. For such applications, a CADD system is often interfaced with a **Computer-Aided Manufacturing (CAM)** system. The CAM system can be programmed to identify and control such manufacturing processes as drilling, boring, turning, milling, and grinding. It also provides for part inspection, testing, and assembly.

The CADD system translates the geometric model of a part or assembly, product information, and necessary engineering data into a computer language such as **Automatically Programmed Tools (APT).** APT produces instructions for numerically controlled (NC), direct NC (DNC), and computer NC (CNC) machine tools.

The latest CAM machine tool language is the (2007) **STEP-NC** machine tool language discussed in more detail in Unit 45. Two different CADD standards are published by (1) the National Council for Advance Manufacturing (NACFAM) and (2) the United States National CAD Standards (NCS).

CARTESIAN COORDINATE SYSTEM

The rectangular coordinate grid of the Cartesian coordinate system provides the foundation on which drawings produced by CAD are generated. The grid is formed by the intersecting of two planes at a right angle. The point of intersection is called the origin, datum, zero reference point, or zero value. The direction and line representing the horizontal plane form the **X** axis, and the vertical plane is the **Y** axis. Dimensions or values to the right of the origin (0) or vertical axis, and above the origin or the horizontal axis, are positive. Dimensions or values to the left of the origin or vertical axis, and below the origin or horizontal axis are negative (Figure 30–1).

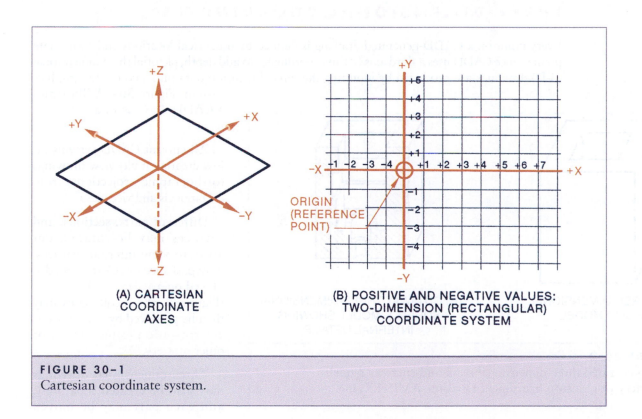

(A) CARTESIAN
COORDINATE
AXES

(B) POSITIVE AND NEGATIVE VALUES:
TWO-DIMENSION (RECTANGULAR)
COORDINATE SYSTEM

FIGURE 30–1
Cartesian coordinate system.

Lines are generated by locating two endpoints in relation to the X and Y axes. The appropriate keyboard command generates the line automatically (Figure 30–2A). Similarly, circles are automatically drawn by locating the center and giving the radius (Figure 30–2B). Circles and arcs may also be generated by locating three points of a circle.

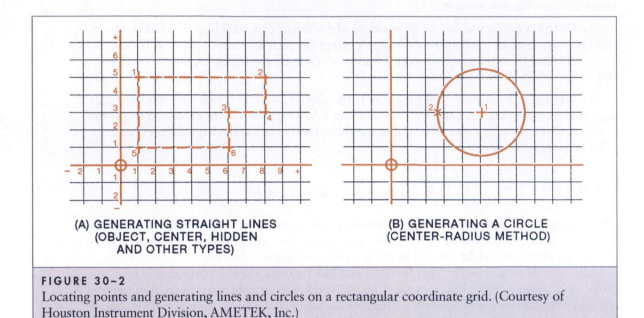

(A) GENERATING STRAIGHT LINES (OBJECT, CENTER, HIDDEN AND OTHER TYPES)

(B) GENERATING A CIRCLE (CENTER-RADIUS METHOD)

FIGURE 30–2
Locating points and generating lines and circles on a rectangular coordinate grid. (Courtesy of Houston Instrument Division, AMETEK, Inc.)

THREE-DIMENSIONAL COORDINATES

Every point on a CADD-generated drawing is defined by numerical locations and coordinate points. Since CADD uses a third axis (Z) and coordinates to add depth, pictorial three-dimensional (3-D) drawings are produced by inputting the three dimension data into a computer graphics system. Figure 30–3A illustrates a CADD drawing of a 3-D solid model.

All current CADD systems allow creating multi-view drawings (orthographic projections) from 3-D models and vice versa.

Different parts, sections, and surfaces may be rotated, cut away to view internal features, drawn, shaded or colored, and displayed on a video display screen (Figure 30–3B). A color copy may then be produced by pen plotters, electrostatic plotters, laser, or other printers.

Three-dimensional drawings may also be created separately, and each part may be moved

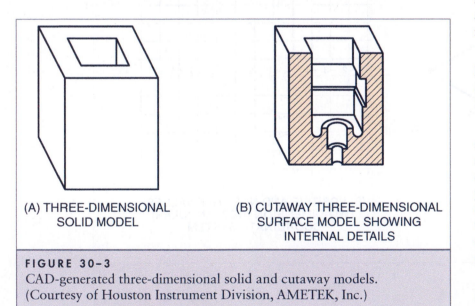

(A) THREE-DIMENSIONAL SOLID MODEL

(B) CUTAWAY THREE-DIMENSIONAL SURFACE MODEL SHOWING INTERNAL DETAILS

FIGURE 30–3
CAD-generated three-dimensional solid and cutaway models. (Courtesy of Houston Instrument Division, AMETEK, Inc.)

into its respective position in an assembly or exploded-parts drawing. Figure 30–4 shows a CADD pictorial assembly drawing produced from separately created drawings of the different parts.

LIBRARY OF SYMBOLS

A typical drawing contains repeated symbols that are produced on conventional drawings by using a template or by rubbing in place selected letters, symbols, or frames from transfer sheets. In CAD, once a symbol is designed, it may be stored, recalled, and reused on a drawing by placing it at the required location.

INPUT AND OUTPUT DEVICES FOR CADD

Information and other graphic instructions are fed into the CADD computer by a number of different devices. Such devices include the common 104-key computer keyboard, a mouse, and a graphics tablet called a digitizer. Frequently, document scanners and specialty pointing devices, such as track balls and 3-D motion control devices, are also found.

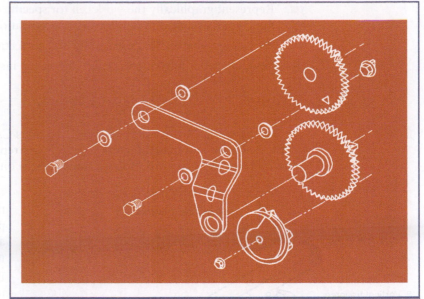

FIGURE 30–4
CAD three-dimensional assembly drawing generated from separately drawn parts. (Courtesy of McDonnell Douglas Manufacturing and Engineering Systems Company.)

- Standard 104-key computer keyboard—an input device for entering text numbers and some basic cursor control.

- Computer mouse—primary input device; used to move the cursor on the screen to access and select options in the drop-down menus of the CADD program. These menus provide lists of commands, symbols, and program control functions.

- Digitizer tablet—used for entering much of the drawing details, such as lines, symbols, circles, and objects; data entered by moving either a stylus or puck over digitizer tablet.
    - Stylus—shaped and held much like a pen; touched to the surface of the tablet at the beginning and end of a line.
    - Puck—shaped like an elongated computer mouse; has a number of buttons at its base for program control. It has a set of cross hairs located in the front of the device that lie over a grid work on the tablet surface. The cross hairs permit more accuracy when digitizing drawings.

- Document scanners that are used in CADD systems are similar to those used in offices but are made bigger in order to handle the large format of "shop" drawings.

The video display screen monitor used with CADD systems are larger than those used on an office computer. Some CADD systems use two video displays for displaying two different views of the same drawing. They can also be used for displaying different documents at the same time. When all information in the display screen (softcopy) is transmitted to the printer, a blueprint (hardcopy) is produced. The hardcopy print is complete with drawing data and other programmed information and can be produced in single or multiple colors.

CADD AND ROBOTICS

CADD systems are used in robotics to perform the following tasks:

- Prepare design and working drawings of robot models.
- Store design and other drawings of parts to be manufactured in the CADD database or produce three-dimensional graphic models.

- Represent graphically the work cell for specific robot functions in a 3-D model including layout, design, and development specifications.
- Provide graphic feedback for study and correction of design and work flow problems.
- Simulate robot movements before the actual installation of a robot in manufacturing.

ROBOTICS: FUNCTIONS

Computer-Integrated Manufacturing (CIM) links CADD/CAM, electronics, fluidics, and robotics to create automated factories. Robots are reprogrammable and have multi-functional moveable mechanical arms. Robots, similar to the one in Figure 30–5, do the following tasks:

- Grip, press, hold, and accurately position workpieces to load and unload parts from work-positioning devices and conveyors (transporters) between machining stations.
- Perform similar processes in loading, unloading, and changing tools and accessories to tool storage drums (matrices) on machine tools.
- Execute programmed manufacturing, assembling, testing, and other processes.

Figure 30–6 provides an example of a computer-activated six-axis robot that is used for handling materials and parts from a conveyor system and loading and unloading a CNC milling center and a CNC turning center in a manufacturing cell. Robots are employed in both handling applications related to the movement of materials and processor applications in manipulating tools and equipment and performing work processes.

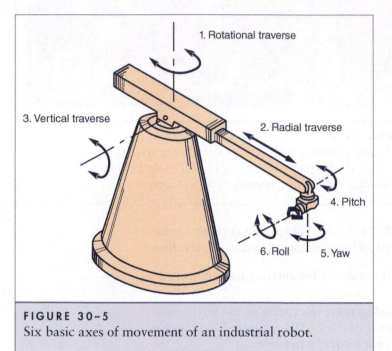

FIGURE 30–5
Six basic axes of movement of an industrial robot.

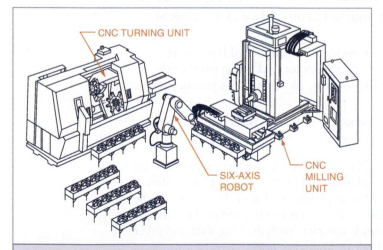

FIGURE 30–6
Computer-controlled six-axis robot handling workpieces between milling and turning machine tools. (Courtesy of Kearney & Trecker Corporation.)

INDUSTRIAL ROBOT DESIGN FEATURES

The three major components of a robot include: (1) the robot hand or end effector, (2 the controller in which program commands are recorded in memory and are programmed to repeat the motions, and (3) the power unit. A brief description of selected robot features follows.

End Effector. A robot generally has two sets of fingers. One set is used to load; the other to unload. After loading and machining, the robot hand turns so the second set of fingers grips and removes the finished part.

Degrees of Freedom. This term identifies the number of axes of motion. For example, eight degrees of freedom means a robot has eight axes of motion.

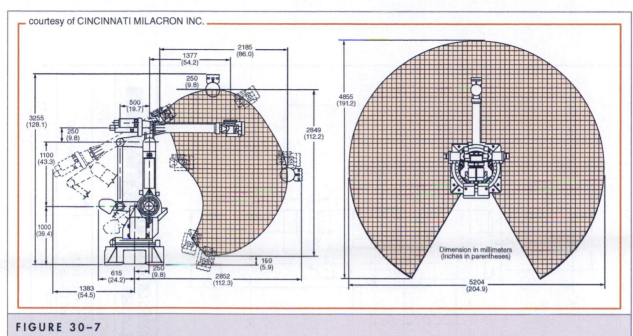

Dimension in millimeters
(Inches in parentheses)

FIGURE 30–7
Robot work envelope describing the range, reach, degrees of freedom, and required floor space.

Operating Reach and Envelope. A CADD graphic 3-D model of a robot performing a specific task includes an inner envelope and an outer envelope (Figure 30–7). Tasks are performed at maximum speed with greater safety and positional accuracy under maximum load within the inner envelope.

Loading and Unloading Design Features. Consideration is given in CADD/CAM systems involving robot applications to the following factors:

- Robot load capacity in terms of weight of the workpiece and grippers (fingers).
- Automatic control of increases or decreases in operating speeds according to work process requirements.
- Safe carrying capacity.
- Clearance of the arm, workpiece work stations, and processes to be performed.
- Positioning repeatability with accuracy.
- Accommodating workpiece irregularities.
- Safe and accurate piloting of a workpiece into a fixture.
- Design of universal hands to accommodate the range of parts within a family of parts.

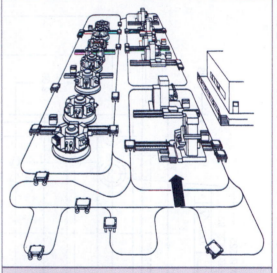

FIGURE 30–8
Model of a computer-directed, random-order, flexible manufacturing system (FMS). Courtesy of Kearney & Trecker Corporation.

Figure 30–8 provides a graphic display of a flexible manufacturing system. An **FMS** interlocks subsystems such as CADD, CAM, computer-directed rail-guided conveyors and robots for handling and for processor applications; turning, milling, grinding, and other CNC machining centers; measurement machine and quality control centers; and other manufacturing cell computer-actuated equipment.

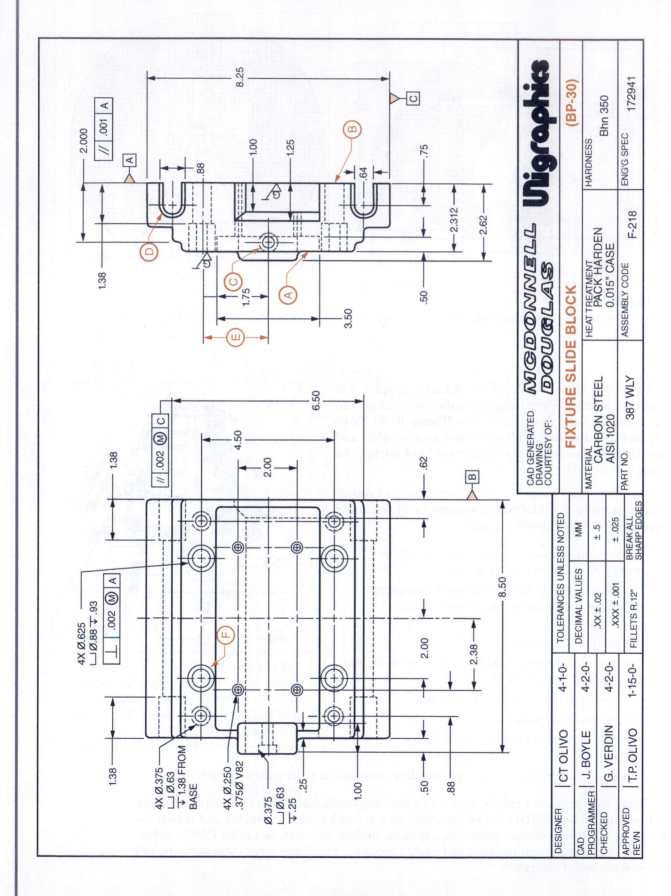

DESIGNER	CT OLIVO	4-1-0-	TOLERANCES UNLESS NOTED		
CAD PROGRAMMER	J. BOYLE	4-2-0-	DECIMAL VALUES	MM	
CHECKED	G. VERDIN	4-2-0-	.XX ± .02	± .5	
			.XXX ± .001	± .025	
APPROVED	T.P. OLIVO	1-15-0-	FILLETS R.12"	BREAK ALL SHARP EDGES	
REVN					

CAD GENERATED DRAWING COURTESY OF: **MCDONNELL DOUGLAS** **Unigraphics**

(BP-30)

FIXTURE SLIDE BLOCK

MATERIAL	HEAT TREATMENT	HARDNESS
CARBON STEEL AISI 1020	PACK HARDEN 0.015" CASE	Bhn 350
PART NO. 387 WLY	ASSEMBLY CODE F-218	ENG'G SPEC 172941

ASSIGNMENT—UNIT 30: FIXTURE SLIDE BLOCK (BP-30)

Student's Name _____

1. Identify the method used to produce the drawing.

2. Give the basic length, height, and thickness of the fixture slide block.

3. Identify the two- and three-place decimal tolerances for unspecified inch and metric dimensions.

4. (a) Identify the material used in the part.
   (b) State the required depth of case hardening.
   (c) Give the Brinell hardness number of the case.

5. (a) Name the machining process for surfaces Ⓐ and Ⓑ.
   (b) Determine the upper and lower decimal inch limits between these surfaces.

6. (a) Write the specifications of the four ⌀ .250" countersunk holes.
   (b) Interpret the meaning of each dimension and symbol.

7. (a) Give the positional (centerline) dimensions for detail Ⓒ.
   (b) Give specifications for this detail.
   (c) Interpret each dimensional value and symbol.

8. (a) List the two basic radii for Ⓓ, in inches.
   (b) State the maximum and minimum depth of Ⓓ, in inches.
   (c) Identify the geometric tolerancing allowance on the center-to-center distance between the slots Ⓓ.
   (d) Translate the meaning of the geometric dimensioning requirements given in the tolerancing frame.

9. Determine the maximum and minimum center-to-center dimension Ⓔ ("/mm).

10. (a) Give complete dimensioning and geometric tolerancing requirements for holes Ⓕ.
    (b) Translate the tolerancing frame information.

11. List three basic CAD system capabilities.

12. State two functions that distinguish CAD from CADD.

13. Identify a major function that is served by interlocking CADD and CAM.

14. (a) Identify the mathematical system that forms the basis for CAD.
    (b) Make a sketch and label the following: (1) point of origin, (2) X axis, (3) Y axis, and (4) positive (+) and negative (−) numbered line values.

15. Provide a general definition of the function of industrial robots.

16. Describe briefly the meaning of the following tasks in relation to robots.
    (a) Simulating robot movements
    (b) Drawing of inner and outer envelopes

1. _____

2. L _____ H _____ T _____

3. XX _____
   XXX _____

4. (a) _____ (b) _____ (c) _____

5. (a) _____
   (b) Upper _____ Lower _____

6. (a) _____
   (b) _____

7. (a) _____
   (b) _____

8. (a) Inch _____ Min. _____
   (b) Max. _____
   (c) _____
   (d) _____

9. (a) Max. _____ Inch
   Min. _____ Inch

10. (a) _____
    (b) _____

11. (1) _____
    (2) _____
    (3) _____

12. (1) _____
    (2) _____

13. _____

14. (a) _____
    (b) _____

15. _____

16. (a) _____
    (b) _____

Speciality Drawings

UNIT

31 Welding Symbols, Representation, and Dimensioning

WELDED STRUCTURES AND SYMBOLS

Welding is an old process that joins metal parts together. The finished welded part will then be able to withstand imposed loads. Many parts today are being fabricated manually or robotically using a welding process.

Complete welding and machining information is conveyed from a part designer to a welder, machinist, or a CNC programmer using graphic symbols. These symbols indicate the required type of weld, specific welding and machining dimensions, and other specifics required to fabricate a part. Standards are established for welding by the American Welding Society (AWS).

TYPES OF WELDED JOINTS

There are five basic types of welded joints specified on drawings. Each type of joint is identified by the position of the parts to be joined together. Parts that are welded by using butt, corner, tee, lap, or edge-type joints are illustrated in Figure 31–1.

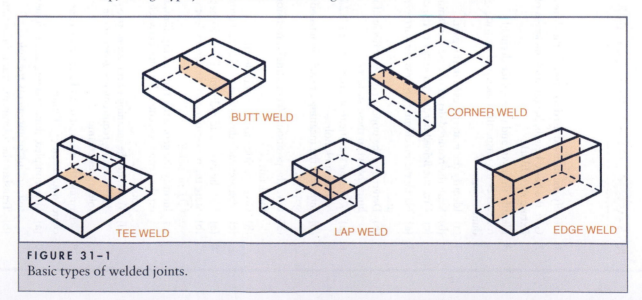

BUTT WELD

CORNER WELD

TEE WELD

LAP WELD

EDGE WELD

FIGURE 31-1
Basic types of welded joints.

AWS SYMBOLS FOR ARC AND GAS WELDS

Two types of welding practices that are commonly used are fusion welding that uses only heat and resistance welding that uses heat and pressure.

The most commonly used arc and gas welds for fusing parts are shown in Figure 31–2. The welds are grouped as follows: ① back or backing, ② fillet, ③ plug and slot, and ④ groove. The groove welds are further identified as (A) square, (B) vee, (C) bevel, (D) U, (E) J, (F) flare, and (G) flare bevel.

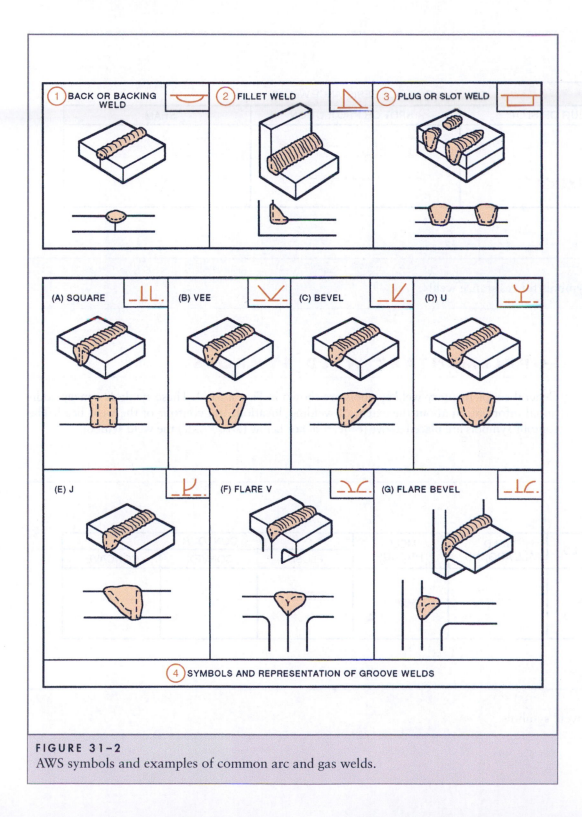

FIGURE 31-2
AWS symbols and examples of common arc and gas welds.

RESISTANCE WELDING: TYPES AND SYMBOLS

Resistance welding is another method of fusing parts. The fusing temperature is produced in the particular area to be welded by applying force and passing electric current between two electrodes and the parts. Resistance welding does not require filler metal or fluxes. The symbols for general types of resistance welds are given in Figure 31–3. These are nonpreferred symbols. Their replacement is recommended by using preferred symbols and including the process reference in the tail.

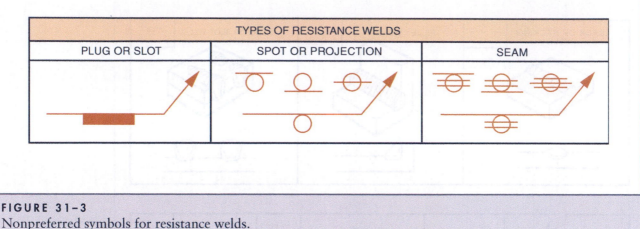

FIGURE 31–3
Nonpreferred symbols for resistance welds.

SUPPLEMENTARY WELD SYMBOLS

General supplementary weld symbols are shown in Figure 31–4. These symbols convey additional information about the extent of welding, location, and contour of the weld bead. The contour symbols are placed above (other side) or below (arrow side) the weld symbol.

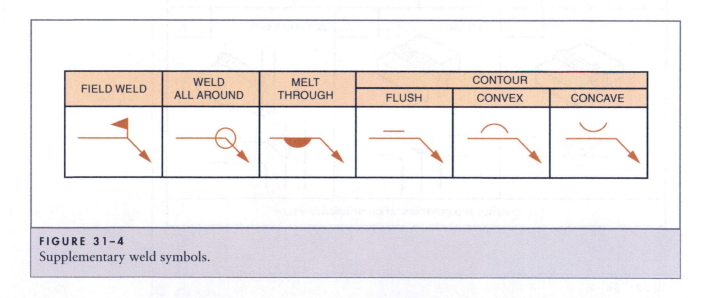

FIGURE 31–4
Supplementary weld symbols.

AMERICAN WELDING SOCIETY STANDARDS FOR SYMBOLS, REPRESENTATION ON DRAWINGS, AND DIMENSIONING

The main parts of the AWS standard welding symbol are identified in Figure 31–5.

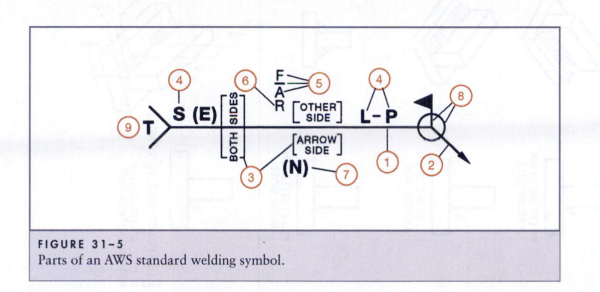

FIGURE 31–5
Parts of an AWS standard welding symbol.

The symbols used for parts (1) through (9) are given in column (A), Figure 31–6. A brief description of the function of each part follows in column (B). Examples of applications appear in column (C). AWS standards are used on many metric drawings, pending the establishment of ISO SI metric standards.

SYMBOL IDENTIFICATION (A)	FUNCTION (B)	EXAMPLES OF REPRESENTATION AND DIMENSIONING (C)
① REFERENCE LINE	BASIS OF WELDING SYMBOL AND ALL ELEMENTS. CONTAINS WELD DATA ABOUT SIZE, TYPE, POSITION, LENGTH, PITCH AND STRENGTH.	FILLET WELD ON ARROW SIDE
② ARROWHEAD	CONNECTS THE REFERENCE LINE TO THE ARROW SIDE MEMBER OF THE JOINT OR GROOVED MEMBER, OR BOTH. POINTS TO THE JOINT WHERE THE WELD IS TO BE PLACED.	FILLET WELD ON OTHER SIDE
③ BASIC WELD SYMBOL AND WELD SIDE	ATTACHED TO REFERENCE LINE TO SHOW THE KIND OF WELD AND THE SIDES TO BE WELDED; OR PROVIDES DETAILED REFERENCE.	FILLET WELD ON BOTH SIDES
④ DIMENSIONS OF WELD	S DENOTES SIZE OR STRENGTH OF CERTAIN WELDS; E EFFECTIVE THROAT OR DEPTH OF PENETRATION; L LENGTH; P PITCH (CENTER-TO-CENTER SPACING OF WELDS), WHERE APPLICABLE. WELD TYPES AND DIMENSIONS ARE PLACED ON OTHER SIDE, ARROW SIDE, OR BOTH SIDES.	$\frac{1}{2}$" LEG FILLET WELD, $1\frac{1}{2}$" LONG, 5" CENTER-TO-CENTER SPACING, OTHER SIDE

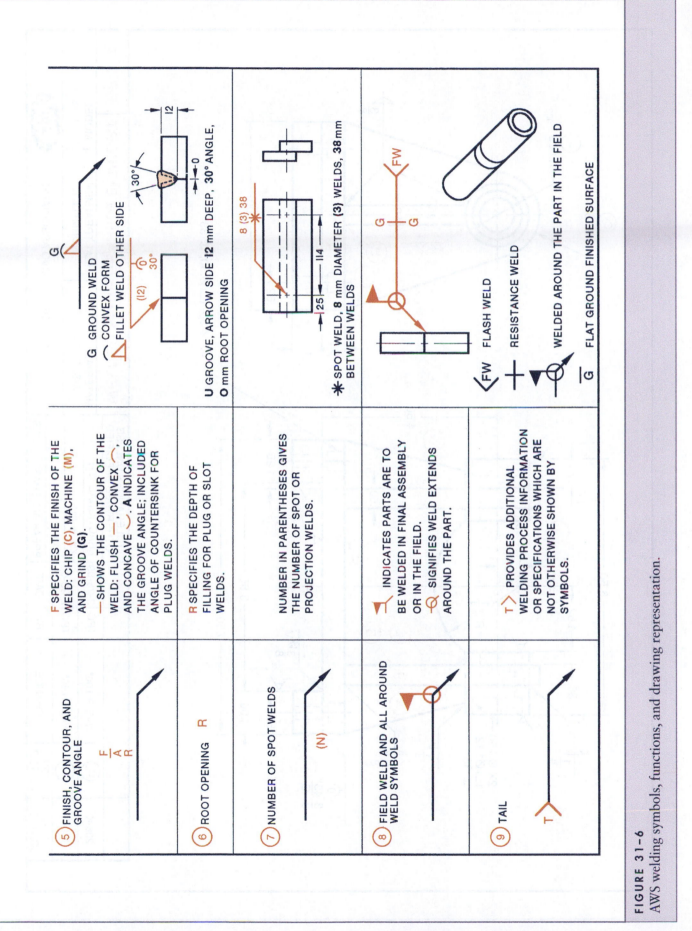

5 FINISH, CONTOUR, AND GROOVE ANGLE

$$\frac{F}{A}\bigg|R$$

F SPECIFIES THE FINISH OF THE WELD: CHIP (C), MACHINE (M), AND GRIND (G).

⌒ SHOWS THE CONTOUR OF THE WELD: FLUSH ⎯, CONVEX ⌢, AND CONCAVE ⌣. A INDICATES THE GROOVE ANGLE; INCLUDED ANGLE OF COUNTERSINK FOR PLUG WELDS.

G GROUND WELD
⌢ CONVEX FORM
FILLET WELD OTHER SIDE

G
30°

U GROOVE, ARROW SIDE **12** mm DEEP, **30°** ANGLE, **0** mm ROOT OPENING

6 ROOT OPENING   R

R SPECIFIES THE DEPTH OF FILLING FOR PLUG OR SLOT WELDS.

7 NUMBER OF SPOT WELDS

(N)

NUMBER IN PARENTHESES GIVES THE NUMBER OF SPOT OR PROJECTION WELDS.

8 (3) 38

✱ SPOT WELD, **8** mm DIAMETER **(3)** WELDS, **38** mm BETWEEN WELDS

8 FIELD WELD AND ALL AROUND WELD SYMBOLS

◤ INDICATES PARTS ARE TO BE WELDED IN FINAL ASSEMBLY OR IN THE FIELD.
○ SIGNIFIES WELD EXTENDS AROUND THE PART.

FW
G   G
FW

◣ FW   FLASH WELD
▼   RESISTANCE WELD
○   WELDED AROUND THE PART IN THE FIELD

9 TAIL

T

T⟩ PROVIDES ADDITIONAL WELDING PROCESS INFORMATION OR SPECIFICATIONS WHICH ARE NOT OTHERWISE SHOWN BY SYMBOLS.

⎯ FLAT GROUND FINISHED SURFACE
G

FIGURE 31-6
AWS welding symbols, functions, and drawing representation.

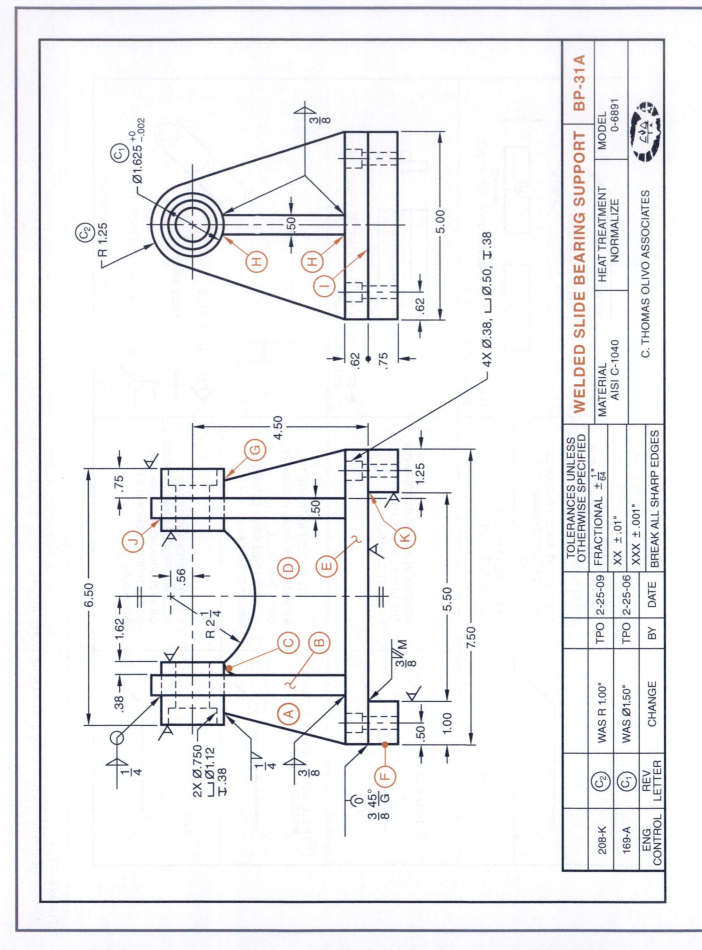

Ø1.625 +0 −.002

R 1.25

.50

5.00

.62

.62

.75

4X Ø.38, ⊔Ø.50, ⊤.38

4.50

.75

6.50

1.62

.56

R 2¼

.38

2X Ø.750 ⊔Ø1.12 ⊤.38

1¼

3/8

.50

.50

1.25

5.50

7.50

1.00

3 45° G

H H I C₁ C₂

J G D E K C B A F

	TOLERANCES UNLESS OTHERWISE SPECIFIED		
	FRACTIONAL ± 1/64"	TPO	2-25-09	WAS R 1.00"	C₂	208-K
	XX ±.01"	TPO	2-25-06	WAS Ø1.50"	C₁	169-A
	XXX ±.001"	BY	DATE	CHANGE	REV LETTER	ENG CONTROL
	BREAK ALL SHARP EDGES		

WELDED SLIDE BEARING SUPPORT BP-31A

MATERIAL AISI C-1040	HEAT TREATMENT NORMALIZE	MODEL 0-6891

C. THOMAS OLIVO ASSOCIATES

ASSIGNMENT A—UNIT 31: WELDED SLIDE BEARING SUPPORT (BP-31A)

Student's Name _____

1. Give the basic dimension for the overall length, depth (width), and height of the bearing support.

2. State what changes were made in dimensions C₁ and C₂ from the original.

3. Determine the upper- and lower-limit dimensions of the shaft bearing diameter.

4. (a) Give the specifications for the C'Bore and recessed holes in the shaft bearings.

    (b) Interpret each symbol and dimension. Include the tolerance for each diameter.

5. Prepare a **materials list** with the following information:

    (a) Give the required number of each part to fabricate the **bearing support**.

    (b) Determine the Nominal size of the rectangular or round bar stock required for each part.

    Note: Convert each two-point decimal size to the equivalent fractional value, to the nearest 1/16".

    (c) Give the code number for the steel to be used in the **bearing support**.

    (d) Identify the heat treatment process.

6. (a) Draw the AWS symbols and values for welds G through K.

    (b) Interpret the meaning of each symbol and give the dimensions for welds G through K.

---

1. L _____ D(W) _____ H _____

2. C₁ _____ C₂ _____

3. Upper _____ Lower _____

4. (a) _____

    (b) _____

5.

Part	Name	No. Req'd (a)	Size (b)
A	End Brace		
B	Upright Bracket		
C	Shaft Bearing		
D	Center Brace		
E	Base		
F	Slide Blocks		

(c) _____ (d) _____

6. G _____ H _____

I _____

J _____

K _____

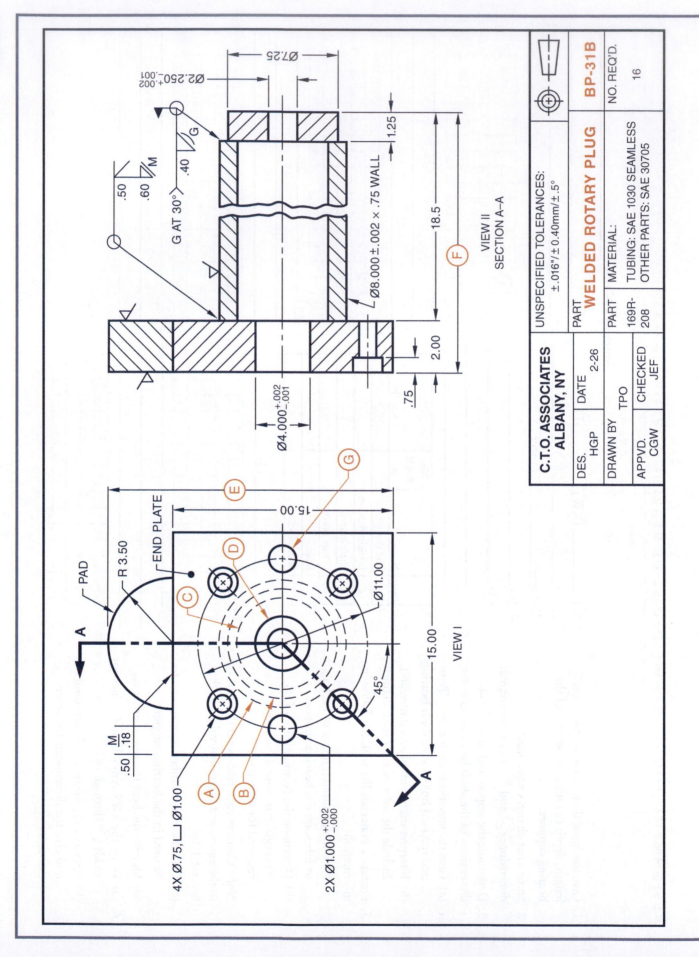

Ø7.25

Ø2.250 +.002 -.001

G AT 30°

.50
.60 M

G

.40

Ø8.000 ±.002 × .75 WALL

1.25

18.5

VIEW II
SECTION A–A

F

2.00

.75

Ø4.000 +.002 -.001

E

END PLATE

15.00

PAD

R 3.50

Ø11.00

45°

15.00

VIEW I

C

D

G

A

B

M .18

.50

4X Ø.75, ⊔ Ø1.00

2X Ø1.000 +.002 -.000

A

A

C.T.O. ASSOCIATES ALBANY, NY	UNSPECIFIED TOLERANCES: ±.016"/±0.40mm/±.5°		
	PART **WELDED ROTARY PLUG**		BP-31B
DES. HGP	DATE 2-26	PART 169R-208	MATERIAL: TUBING: SAE 1030 SEAMLESS OTHER PARTS: SAE 30705
DRAWN BY TPO			
APPVD. CGW	CHECKED JEF		NO. REQ'D. 16

ASSIGNMENT B—UNIT 31: WELDED ROTARY PLUG (BP-31B)

Student's Name _____

NOTE: All dimensions are to be given in both inch and millimeter values.

1. Name the two views.

2. List (a) the wall thickness of the tubing and (b) the material for the tube body and other parts.

3. Give the basic diameters of Ⓐ, Ⓑ, and Ⓒ.

4. Determine the maximum and minimum diameters for the bored hole Ⓓ.

5. Establish the maximum overall height Ⓔ and length Ⓕ.

6. Give the limits for the Ⓖ holes.

7. Identify (a) the minimum diameter and (b) the depth of the counterbored holes.

8. Specify the basic size of the welded pad.

9. Determine how many surfaces are to be machined.

10. Give the maximum dimensions of the circular end plate.

11. Indicate (a) the type of welded joint between the pad and the 2.00″ thick-plate and (b) give the weld specifications.

12. Interpret the weld specifications for the tubing and square end plate.

13. Draw the AWS symbols that describe the weld for the circular end plate and tubing body.

14. Interpret the symbols and dimensions for the circular end plate and body weld.

1. VIEW I _____ VIEW II _____

2. (a) Wall thickness _____

   (b) Materials _____

3. Ⓐ _____ basic diameter

   Ⓑ _____ basic diameter

   Ⓒ _____ basic diameter

4. Ⓓ Maximum diameter _____

      Minimum diameter _____

5. Ⓔ Height _____ Ⓕ Length _____

6. UL IN _____ UL MM _____

   LL IN _____ LL MM _____

7. (a) Minimum diameter _____

   (b) Minimum depth _____

8. _____

9. _____

10. _____

11. (a) Type _____

    (b) Specifications _____

12. _____

13.

14. _____

32

Surface Development and Precision Sheet Metal Drawing

In major industries, such as construction, HVAC, automotive, aerospace, electronic, and farm equipment, many parts are produced from sheets of metal, plastic, carbon fiber material, medium-density fiberboard (MDF), and similar materials that can be cut and formed. Ducts, pipes, holding tanks, cabinets, and other sheet material objects consist of sides and surfaces that are square, cylindrical, prism-or cone-shaped, or a combination. The surfaces may also intersect at a specified angle.

Each form may be produced by first laying out each section or area in full size as a single form on a flat, plane surface. This technique of layout is known as surface development. The part as produced may be called a template, a stretchout, or a pattern.

The stretch-out is then cut to size by sawing, stamping, shearing, chemical milling, ultrasonic, electron beams, or lasers. Depending on the complexity of the pattern, it may be shaped by folding, bending, rolling, or beading. Surfaces with complicated curvatures are produced by drop hammer, fluidic presses, stretch presses, die stamping, vacuum forming, compression molding, etc.

After the stretch-outs have been bent into final form, matching surfaces may need to be held together. Fastening choices include adhesives, chemical bonding, brazing, welding, assorted screws, rivets, rolling, bending, and clinching. All the steps from layout to product completion can be accomplished manually or by using CAD/CIM techniques or some of both methods.

PATTERN DEVELOPMENT PROCESSES

A flat-form pattern in sheet metal work may be developed by parallel line development, radial line development, or triangulation. Sheet metal (or nonmetallic) parts that connect pipes or openings of different sizes and shapes are known as transition pieces. Two common transition pieces are shown in Figure 32–1.

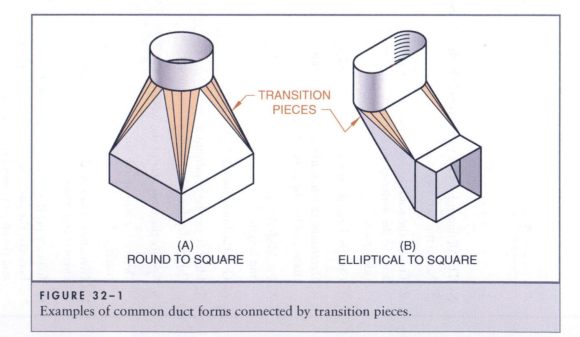

TRANSITION PIECES

(A)
ROUND TO SQUARE

(B)
ELLIPTICAL TO SQUARE

FIGURE 32–1
Examples of common duct forms connected by transition pieces.

Parallel Line Development

Parallel line development means that the pattern is drawn full size by using parallel lines for the width, height, and depth features of an object. If the object is cylindrical and it is cut by a plane at an angle (truncated), the pattern is developed by dividing the circle of the cylinder into an equal number of parts (elements). The height of the truncated cylinder at each element is transferred to the stretchout. Figure 32–2 shows the technique of developing a pattern for a truncated cylinder.

Using the same method, it is possible to develop the pattern or template for two, three, or more pieces of the same diameter (like an elbow) that are to fit together.

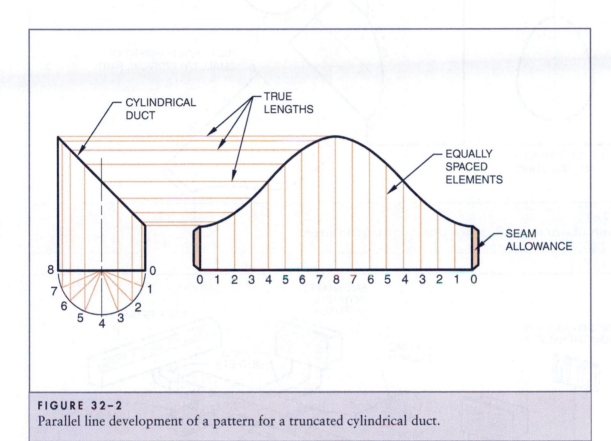

FIGURE 32–2
Parallel line development of a pattern for a truncated cylindrical duct.

Intersections

The point or place where the several parts of a sheet metal object join is known as the line of intersection or cutting plane line. The line may be straight, as in the case of two equal square ducts that intersect. In other instances, like a square duct being cut by a cylindrical duct, the line of intersection or the hole where the separate ducts join is respresented by curved lines on both pieces.

Intersections are developed by transferring true length heights from successive elements onto a plane surface layout for each part. The principle used in developing intersections is illustrated in Figure 32–3. While the example shows two different size cylinders intersecting at an angle, the same technique may be used with intersecting prisms, cylinders, cones, and other forms.

Major components in a heating and cooling air conditioning system and the sheet metal duct work for a typical installation are shown in Figure 32–4.

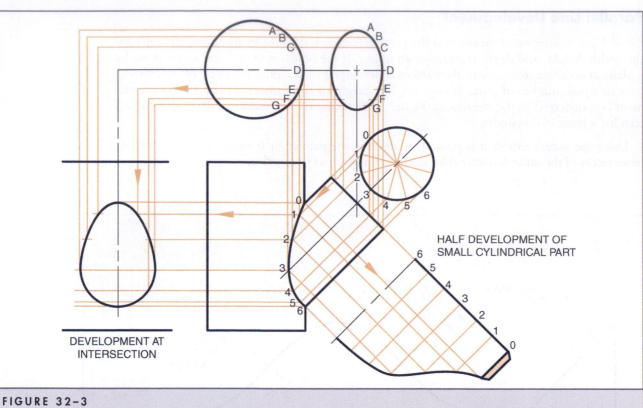

A
B
C
D
E
F
G

A
B
C
D
E
F
G

0
1
2
3
4
5
6

0
1
2
3
4
5
6

HALF DEVELOPMENT OF
SMALL CYLINDRICAL PART

6
5
4
3
2
1
0

0
1
2
3
4
5
6

DEVELOPMENT AT
INTERSECTION

FIGURE 32–3
Developments of two cylindrical parts that intersect at an angle.

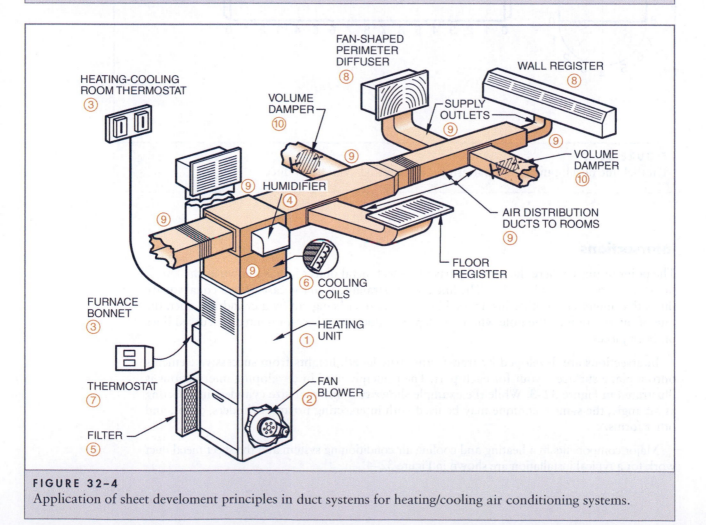

FAN-SHAPED
PERIMETER
DIFFUSER
⑧

WALL REGISTER
⑧

HEATING-COOLING
ROOM THERMOSTAT
③

VOLUME
DAMPER
⑩

SUPPLY
OUTLETS

⑨

⑨

⑨

⑨

VOLUME
DAMPER
⑩

⑨ HUMIDIFIER
④

⑨

AIR DISTRIBUTION
DUCTS TO ROOMS
⑨

⑨

FLOOR
REGISTER

FURNACE
BONNET
③

⑨

COOLING
COILS
⑥

HEATING
UNIT
①

THERMOSTAT
⑦

FAN
BLOWER
②

FILTER
⑤

FIGURE 32–4
Application of sheet develment principles in duct systems for heating/cooling air conditioning systems.

BEND, SEAM, AND LAP ALLOWANCES

Allowances, or the amount of extra material required, must be provided on layouts for bends, seams, and laps. It is critical to include such allowances on drawings to ensure that the desired final dimension is achieved during actual production. The bend allowance varies based on the sheet material, the thickness of the materials, and the form and size of the bend. Although there is a specific formula used to calculate the bend allowance, charts providing standardized information about allowances used in different industries can be found in sources such as Machinery's Handbook.

Bend allowances for straight, right-angle, and open or closed bends appear on drawings as a dimension. Drawings of precision sheet metal parts usually show the bend line for the starting of the bend and a dimension that gives the allowance for the bend.

Drawing developments for sheet metal ducts and pipes make provision for wiring, seaming, hemming, and other processes for joining the ends together and locking them in position (Figure 32–5).

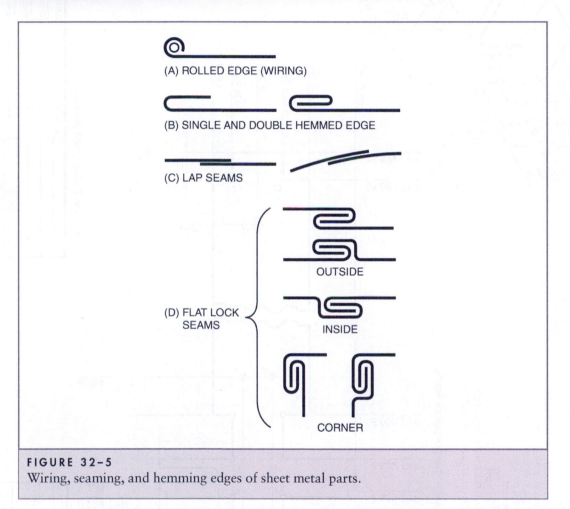

(A) ROLLED EDGE (WIRING)

(B) SINGLE AND DOUBLE HEMMED EDGE

(C) LAP SEAMS

OUTSIDE

(D) FLAT LOCK SEAMS

INSIDE

CORNER

FIGURE 32–5
Wiring, seaming, and hemming edges of sheet metal parts.

PRECISION SHEET METAL PARTS PRODUCTION

The increasing number of applications of thingage metals for parts within instruments, gages, electronic, and other precision equipment requires layouts and parts that are produced to precise machine shop standards. Such sheet metal parts are often blanked, perforated, formed, and trimmed using punches and dies.

In other instances, parts may be manually or computer-programmed and produced on numerical control machines. Ordinate drawings are commonly used for producing precision sheet metal parts. Ordinate dimensions are referenced to zero datums to provide position and size dimensions for production of the part of NC or CNC equipment.

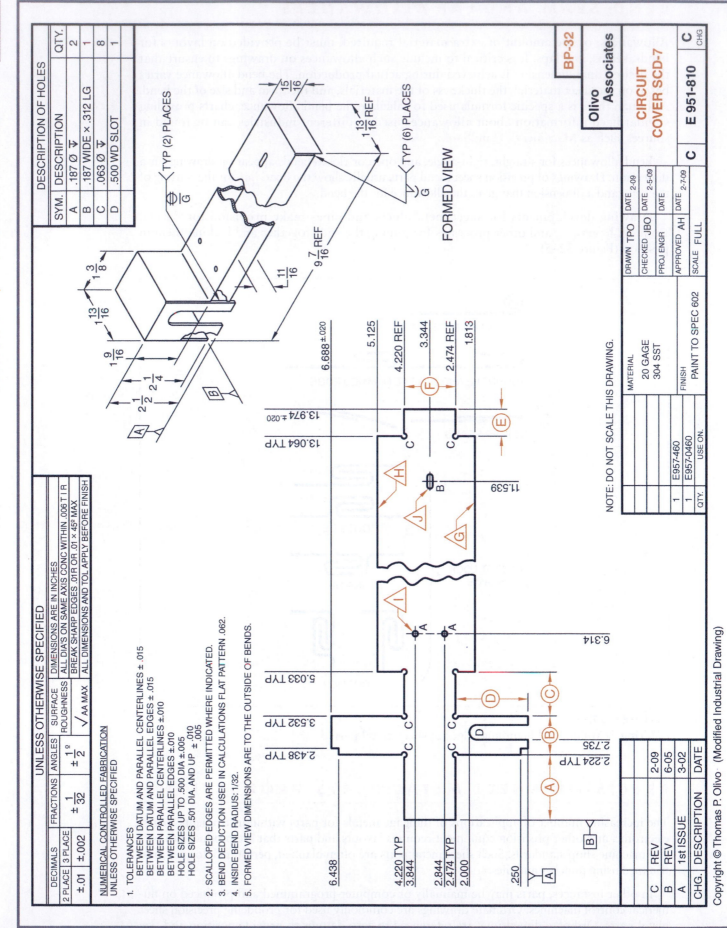

FORMED VIEW

SYM.	DESCRIPTION OF HOLES DESCRIPTION	QTY.
A	.187 Ø ▽	2
B	.187 WIDE × .312 LG	1
C	.063 Ø ▽	8
D	.500 WD SLOT	1

UNLESS OTHERWISE SPECIFIED

DECIMALS		FRACTIONS	ANGLES	SURFACE ROUGHNESS
2 PLACE	3 PLACE	± 1/32	± 1°/2	√ AA MAX
±.01	±.002			

DIMENSIONS ARE IN INCHES
ALL DIAS ON SAME AXIS CONC WITHIN .006 T I R
BREAK SHARP EDGES .01R OR .01 × 45° MAX
ALL DIMENSIONS AND TOL APPLY BEFORE FINISH

NUMERICAL CONTROLLED FABRICATION
UNLESS OTHERWISE SPECIFIED

1. TOLERANCES
   BETWEEN DATUM AND PARALLEL CENTERLINES ± .015
   BETWEEN DATUM AND PARALLEL EDGES ± .015
   BETWEEN PARALLEL CENTERLINES ±.010
   BETWEEN PARALLEL EDGES ±.010
   HOLE SIZES UP TO .500 DIA ±.005
   HOLE SIZES .501 DIA. AND UP ± .005

2. SCALLOPED EDGES ARE PERMITTED WHERE INDICATED.

3. BEND DEDUCTION USED IN CALCULATIONS FLAT PATTERN .062.

4. INSIDE BEND RADIUS: 1/32.

5. FORMED VIEW DIMENSIONS ARE TO THE OUTSIDE OF BENDS.

NOTE: DO NOT SCALE THIS DRAWING.

			DRAWN TPO	DATE 2-09
			CHECKED JBO	DATE 2-5-09
			PROJ ENGR	DATE
			APPROVED AH	DATE 2-7-09

MATERIAL: 20 GAGE 304 SST
FINISH: PAINT TO SPEC 602
SCALE FULL

QTY.		USE ON.
1	E957-460	
1	E957-0460	

Olivo Associates

CIRCUIT COVER SCD-7

BP-32

C | E 951-810 | C
CHG

CHG.	DESCRIPTION	DATE
C	REV	2-09
B	REV	6-05
A	1st ISSUE	3-02

ASSIGNMENT—UNIT 32: CIRCUIT-DIAL COVER SCD 7 (BP-32)

Student's Name _____

1. State the type of dimensioning that is used for the (a) pictorial (formed view) sketch and (b) the layout drawing.

2. Give the specifications for (a) the material used for the Circuit-Dial Cover and (b) the required finish.

3. Identify the connectivity tolerance (unless otherwise specified) for all diameters on the same axis.

4. Specify the kind of fabrication equipment that is to be used in producing the part.

5. Provide the following information (unless otherwise specified):
   (a) Inside bend radius
   (b) Bend allowance for laying out the flat pattern

6. Give the following general tolerances:
   (a) Hole diameters up to 0.500"
   (b) Between parallel center lines
   (c) Between a datum and parallel center lines
   (d) Between parallel edges

7. Indicate the dimensions to which sharp edges are to be broken.

8. (a) Show how the two different types of welds are represented.
   (b) Interpret the meaning of each specification.

9. Establish the maximum and minimum dimensions for each of the following features:

   (c) Location of the dimensions in the formed view

   (a) Elongated hole B
   (b) Slot D
   (c) Holes A
   (d) Corner radius C
   (e) Length and Height of the flat pattern from the datums

10. Identify the REFERENCE dimensions for the flat pattern and the formed view.

11. Establish the following basic dimensions: Ⓐ through Ⓕ (including the bend allowances) of the finish-formed part. Give the dimensions in fractional inch values.

12. Compute the maximum and minimum dimensions between the following features:
    (a) Surfaces ◁G▷ and ◁H▷
    (b) Datum ◻B◻ and center line ◁I▷
    (c) Center lines ◁I▷ and ◁J▷

---

1. (a) _____
   (b) _____

2. (a) Material _____
   (b) Finish _____

3. _____

4. _____

5. (a) _____ (b) _____
   (c) _____

6. (a) _____ (b) _____
   (c) _____ (d) _____

7. _____

8. (a) (1) _____ (2) _____
   (b) _____

9. (a) Width (DIA) _____   Maximum _____ Minimum _____
       Length _____
   (b) Height _____
   (c) _____
   (d) _____

9. (e) Length _____
       Height _____

10. (a) Flat pattern _____
    (b) Formed view _____

11. Ⓐ _____
    Ⓑ _____
    Ⓒ _____
    Ⓓ _____
    Ⓔ _____
    Ⓕ _____

12. (a) _____   Maximum _____ Minimum _____
    (b) _____
    (c) _____

Working Drawings

33

Detail Drawings and Assembly Drawings

A **working drawing** is one that includes all the information necessary to successfully and accurately complete a job. A **detail drawing** is a working drawing that includes a great deal of data, including the size and shape of the project, what kinds of materials should be used, how the finishing should be done, and what degree of accuracy is needed. Every detail must be given. An **assembly drawing,** on the other hand, may have very little detail. The purpose of this type of technical drawing is to depict how the machine is to be put together (assembled). See Figure 33–1.

A well-prepared working drawing will follow the style and practices of the office or industry where it is to be used, but there are certain rules that must be followed. For instance, to ensure sharp contrast, proper line techniques must be used; dimensions and notes must be accurate and easy to read; and standard terms and abbreviations must be used. When the drawing is complete, it should be very carefully checked by a second person as a precaution against error.

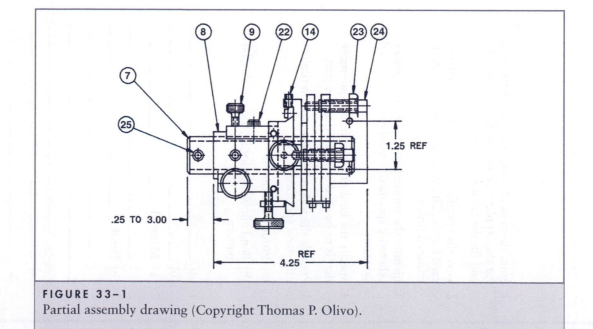

FIGURE 33–1
Partial assembly drawing (Copyright Thomas P. Olivo).

DETAIL DRAWINGS

Detail drawings provide all information necessary to manufacture a part. Typically, only one part is detailed in each drawing; however, there may be multiple views of that part. In some cases, though, several parts of a machine may be detailed on one sheet. Sometimes, a separate drawing is made for each of several departments, such as pattern-making, welding, or machining. A drawing of this type will contain only dimensions and information needed by the department for which it is made.

In the language of technical drawing, an object is drawn giving its shape and its size. This is true whether it is a simple machine, a steam or gas engine, a building, an automobile, or a satellite. The only part that may not need to be drawn is a standard part, one which can probably be bought from an outside supplier more economically than it can be manufactured. These parts include screws, keys, pins, and bolts. They do not need to be drawn, but they do need to be included in the general information that is part of each sheet.

All or some of the following items should be included in detail drawings.

1. Multiple views, as necessary
2. Material(s) used to make the part
3. Dimensions
4. General notes and specific manufacturing information
5. Identification of the project name, the part, and the part number
6. Name or initials of who worked on or with the drawing
7. Any engineering changes and related information

Usually, the detail drawings have information that is classified into three groups:

1. Shape description that describes or shows the shape of the part
2. Size description that shows the size and location of features of the part
3. Specifications regarding items such as material, finish, and heat treatment

ASSEMBLY DRAWINGS

A very large number of products contain more than one part to them. An assembly drawing identifies each part and shows how they all fit together. The value of assembly drawings is that they show how the parts fit together, the overall look of the construction, and dimensions needed for installation. Hidden lines and unnecessary details are often omitted. In some situations, sectional or auxiliary views are given.

When using CAD, individual parts (details) can be merged together to create an assembly or a working drawing before the parts are actually made. Three-dimensional (3-D) models can be created in such a manner that it is possible to superimpose images and to graphically measure clearances. If the parts have been designed or drawn incorrectly, the errors will frequently be obvious and appropriate correction can be made. All these details make the final print accurate and the resultant parts function correctly.

There are actually a variety of assembly drawings. Some are:

1. Layout assembly drawings used as a first step in developing a new product.
2. Installation assembly drawings are necessary to show how to install large pieces of equipment.
3. Diagram assembly drawings use conventional symbols to show the approximate location and/or sequence of parts to be assembled.
4. Working assembly drawings are fully dimensioned and noted. For very simple products, they can be used as detail drawings.
5. Exploded assembly drawings pictorially show parts laid out in their correct order of assembly as if they had been exploded (Figure 33–2).

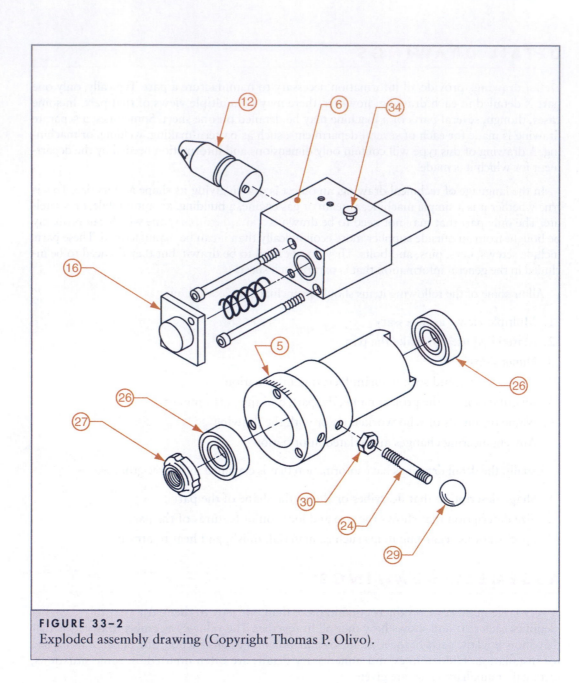

FIGURE 33-2
Exploded assembly drawing (Copyright Thomas P. Olivo).

TITLE BLOCK

This part of a drawing is an outlined rectangular space in the lower right-hand corner of the drawing or across the bottom (Figure 33–3A). The latter is referred to as a record strip. This box contains a variety of information including:

1. The title of the drawing
2. The name of the company and its location
3. The name of the machine or unit of which the drawing is a subassembly
4. The number of the drawing
5. The part number
6. The number of parts to be made
7. The material from which the part is to be made
8. Drafting form record (who drew, checked, and approved the drawing, including the date of each)

9. Tolerances

10. Finishes

11. Heat treatment

12. Scale

13. Angle of projection (1st or 3rd angle)

14. Geometric drawing symbols used on the drawing

15. The sheet number (1 of 2, 2 of 2, etc.)

16. The date on which it was drawn

17. The weight of the part

ITEM	QTY	PART NUMBER	DESCRIPTION
25	2	DDS-A2-24	SCREW, SET, NYLON PT NORDEX
24	2		SOC. HD. CAPSCR, #10-32X 1⅜L6 ST. ST.
23	2	77A025	JACK SCREW
22	1	77A024	DOG PT SET SCREW
21	1	77B011	SLIDE
20	1	DFS-A1-6	SOC SET SCREW NORDEX
19	4	DKS-A3-5	SOC HD. SCREW (4-40X⅜) NORDEX
18	3	EPS-E1-9	PIN NORDEX
17	20	DKS-A3-3	SOC HD. SCREW (4-40X¼) NORDEX
16	2	BAS-A1-4	COLLAR
15	4		WASHER, FLAT #4 ST. ST.
14	4	DFS-A1-8	SOC SET SCREW (4-40X¼) NORDEX
13	1	77A018	GEAR
12	2	AGB-A2-4	BEARING NORDEX
11	1	77A023	THUMB SCREW
10	2	77A022	THUMB SCREW
9	1	77A021	THUMB SCREW
8	1	77B009	HOUSING
7	1	77B012	RACK SLEEVE
6	2	77A019	SPRING
5	2	77A020	GIB
4	1	77B008	PLATE BRACKET
3	1	77B010	SLIDE
2	1	77A017	PLATE
1	REF	AS SPEC	SHAFT TYPE SI SPERRY PROD.

PARTS LIST

AUTOMATION INDUSTRIES, INC.
SPERRY DIVISION

REVISIONS			**MANIPULATOR ASSEMBLY**	P-42
NO.	DATE	BY		
1	7/10/06			
2	3/12/07		DRAWN BY SCALE MATERIAL	
3	8/26/08		CHKD DATE DRAWING NO.	
4			FILED APPD **77D025**	
5				

FIGURE 33-3

(A) Title block, (B) Parts list, (C) Revision block (Copyright Thomas P. Olivo).

PARTS LIST (BILL OF MATERIALS)

The graphically depicted parts of the product are keyed to a parts list with reference numbers. This parts list usually appears right above the title block. It will include the different parts that go into the assembly shown on the drawing, the description and name of each part, the quantity needed of each part, and any materials specifications. The parts should be listed in general order of size or importance. If parts are to be purchased from a vendor, the name or an identification code should be listed.

If a drawing is complicated, containing many parts, the parts list may be on a separate piece of paper that is attached to the drawing. (See Figure 33–3B.)

REVISIONS BLOCK

Sometimes, after a drawing has been released to a production department, it is necessary to make design revisions. This may be for design improvement, errors found in the original drawing, production problems, or other reasons. These changes must be approved and shown on the drawing in what is known as a revisions or change block. It is usually located to the left or above the title block. Typically, the entry includes a brief description of the change and an identifying letter referencing it to the specific location on the drawing and the date. The initials of the draftsperson making the change and those approving it are required. (See Figure 33–3C.)

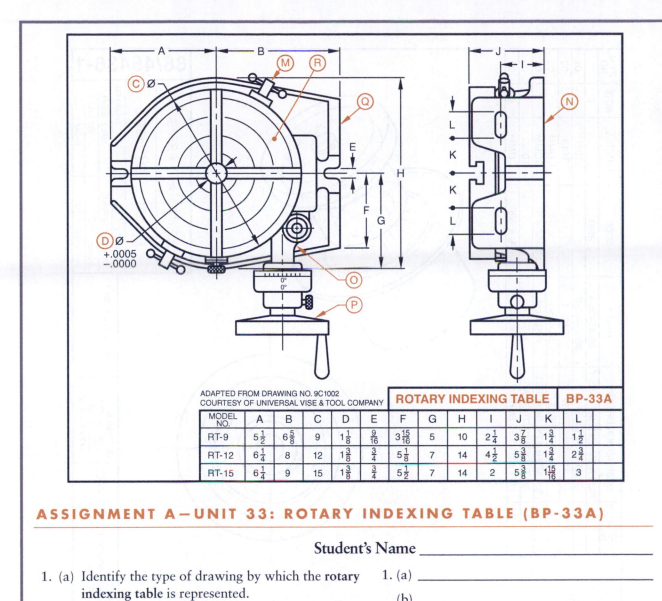

ADAPTED FROM DRAWING NO. 9C1002
COURTESY OF UNIVERSAL VISE & TOOL COMPANY

ROTARY INDEXING TABLE												BP-33A
MODEL NO.	A	B	C	D	E	F	G	H	I	J	K	L
RT-9	$5\frac{1}{2}$	$6\frac{5}{8}$	9	$1\frac{1}{8}$	$\frac{9}{16}$	$3\frac{15}{16}$	5	10	$2\frac{1}{4}$	$3\frac{7}{8}$	$1\frac{3}{4}$	$1\frac{1}{2}$
RT-12	$6\frac{1}{4}$	8	12	$1\frac{3}{8}$	$\frac{3}{4}$	$5\frac{1}{8}$	7	14	$4\frac{1}{2}$	$5\frac{3}{8}$	$1\frac{3}{4}$	$2\frac{3}{4}$
RT-15	$6\frac{1}{4}$	9	15	$1\frac{3}{8}$	$\frac{3}{4}$	$5\frac{1}{2}$	7	14	2	$5\frac{3}{8}$	$1\frac{15}{16}$	3

ASSIGNMENT A—UNIT 33: ROTARY INDEXING TABLE (BP-33A)

Student's Name _____

1. (a) Identify the type of drawing by which the **rotary indexing table** is represented.

   (b) Name the two views.

2. State two important functions that are served by this drawing.

3. Provide model numbers for which specifications are given.

4. Identify the letter on the drawing that represents each of the following parts or features:

   (a) **rotary table plate**   (e) **indexing crank**

   (b) **base**   (f) **gear reduction**

   (c) **right-angle plate**   **unit**

   (d) **table locking screws**

5. Use the dimensional information given in the drawing specifications. Record the sizes as indicated in the table for each specified item as follows:

   (a) RT-9 (A), (B), (C), and (D).

   (b) RT-12 (E), (F), (G), and (H).

   (c) RT-15 (I), (J), (K), and (L).

1. (a) _____
   (b) _____
2. (a) _____
   _____
   _____
   (b) _____
   _____
   _____
3. Models _____
4. (a) ◯   (c) ◯   (e) ◯
   (b) ◯   (d) ◯   (f) ◯
5. (a) A = ____   (b) E = ____   (c) I = ____
   B = ____   F = ____   J = ____
   C = ____   G = ____   K = ____
   D = ____   H = ____   L = ____

36/45436-1

TILLER SHAFT ASSEMBLY

BP 33B   SH 1 OF 5

drawn by ELB	11MAY09
checked by Kerri	HALF SCALE
super TPO	36° TILLER
released CTO	order # 15576

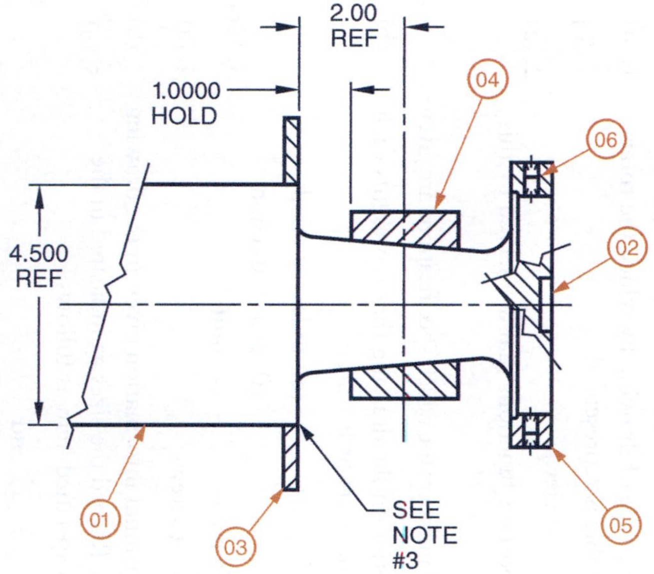

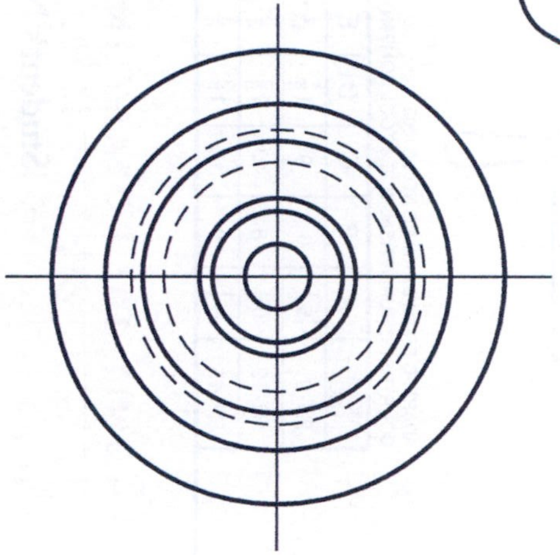

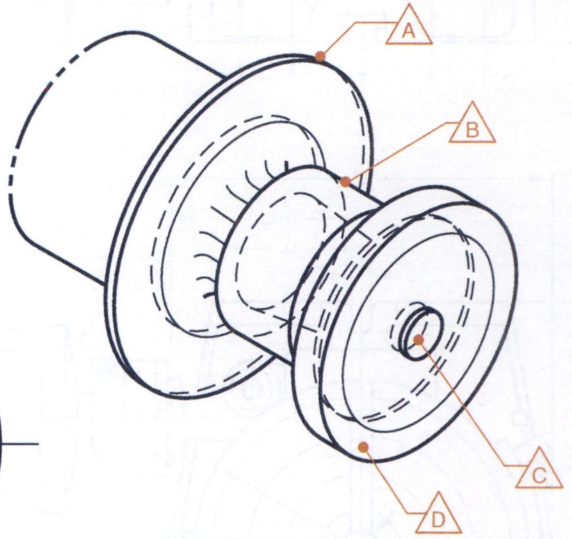

NOTE:
NON SPECIFIED TOLERANCE
∢ ± .05°

DECIMALS:
.XX = .02±
.XXX = .001±
.XXXX = .0005±

FRACTIONS = 1/64 "±

Parts List

	QUANTITY		DESCRIPTION OF PART	PART #	CODE	MATERIAL	NOTE	ITEM NO.
GR-2	702	GR-1						
		1	TILLER TAPER SHAFT ASS'Y	36/45436-1	ANG		2	01
		1	TILLER TAPER SHAFT	36/45436-01		T001	1	02
		1	HARD BUTTON	15415-1		T013		03
		1	SLIPON RING	SL455-1	7" × 1/4"	COMM		04
		1	5° TAPER 4 PC BEARING	WAU6std	3 1/2" × 2"	COMM		05
		1	30 TOOTH CAST GEAR	h-302-1		T0322		06
		2	INT HEX SET SCREW	twf998	1/4" × 3/8"	COMM	4	

36/45436-1

REF. 36/45126   NOTES:

1. HEAT ITEM #01 TO 300°, CHILL ITEM #02 TO 200°.
   PRESS GENTLY, DO NOT HAMMER IN PLACE

2. WIRE TAG WITH ASSEMBLY PART #

3. .002 INTERFERENCE PRESS FIT

4. USE OIL PROOF THREAD LOC

2.00 REF

1.0000 HOLD

4.500 REF

SEE NOTE #3

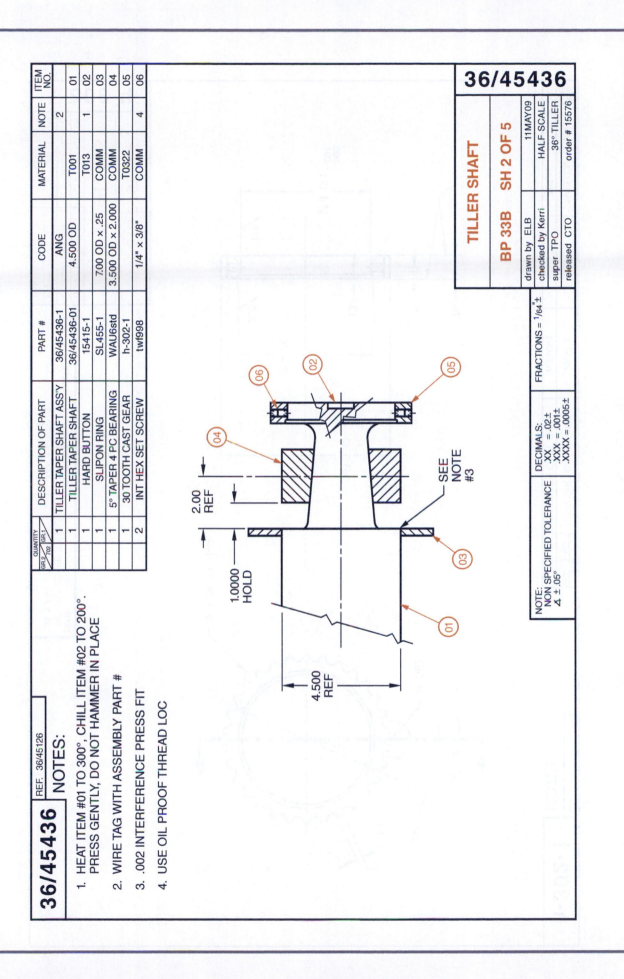

36/45436

REF. 36/45126 NOTES:

1. HEAT ITEM #01 TO 300°, CHILL ITEM #02 TO 200°. PRESS GENTLY, DO NOT HAMMER IN PLACE

2. WIRE TAG WITH ASSEMBLY PART #

3. .002 INTERFERENCE PRESS FIT

4. USE OIL PROOF THREAD LOC

QUANTITY		DESCRIPTION OF PART	PART #	CODE	MATERIAL	NOTE	ITEM NO.
GR 2 702	GR 1						
	1	TILLER TAPER SHAFT ASS'Y	36/45436-1	ANG		2	01
	1	TILLER TAPER SHAFT	36/45436-01	4.500 OD	T001		01
	1	HARD BUTTON	15415-1		T013	1	02
	1	SLIPON RING	SL455-1	7.00 OD x .25	COMM		03
	1	5° TAPER 4 PC BEARING	WAU6std	3.500 OD x 2.000	COMM		04
	1	30 TOOTH CAST GEAR	h-302-1		T0322		05
	2	INT HEX SET SCREW	twf998	1/4" x 3/8"	COMM	4	06

36/45436

TILLER SHAFT

BP 33B   SH 2 OF 5

drawn by ELB	11MAY09
checked by Kerri	HALF SCALE
super TPO	36° TILLER
released CTO	order # 15576

FRACTIONS = $^1/_{64}$"±

DECIMALS:
.XX = .02±
.XXX = .001±
.XXXX = .0005±

NOTE:
NON SPECIFIED TOLERANCE
∡ ± .05°

2.00 REF

1.0000 HOLD

4.500 REF

SEE NOTE #3

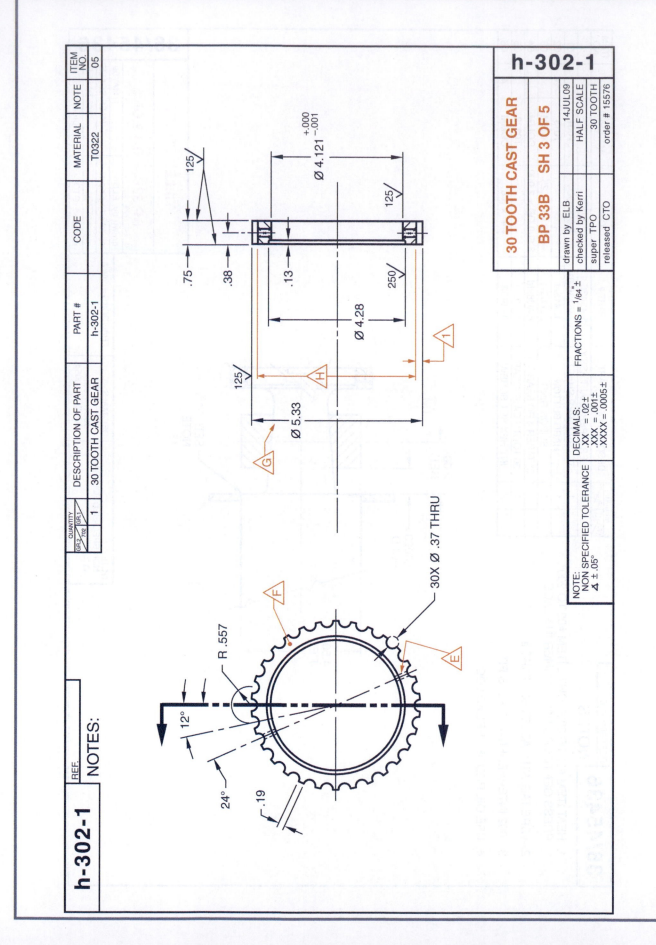

h-302-1

NOTES:

REF.

	QUANTITY		DESCRIPTION OF PART	PART #	CODE	MATERIAL	NOTE	ITEM NO.
GR 2	702	GR 1						
		1	30 TOOTH CAST GEAR	h-302-1		T0322		05

h-302-1

30 TOOTH CAST GEAR

BP 33B   SH 3 OF 5

drawn by ELB	14 JUL 09
checked by Kerri	HALF SCALE
super TPO	30 TOOTH
released CTO	order # 15576

DECIMALS:
.XX   = .02±
.XXX = .001±
.XXXX = .0005±

FRACTIONS = ¹/₆₄"±

NOTE:
NON SPECIFIED TOLERANCE
∡ ±.05°

Ø 4.121 $^{+.000}_{-.001}$

Ø 4.28

Ø 5.33

.75

.38

.13

125

125

125

250

G

H

1

R .557

12°

24°

.19

30X Ø .37 THRU

F

E

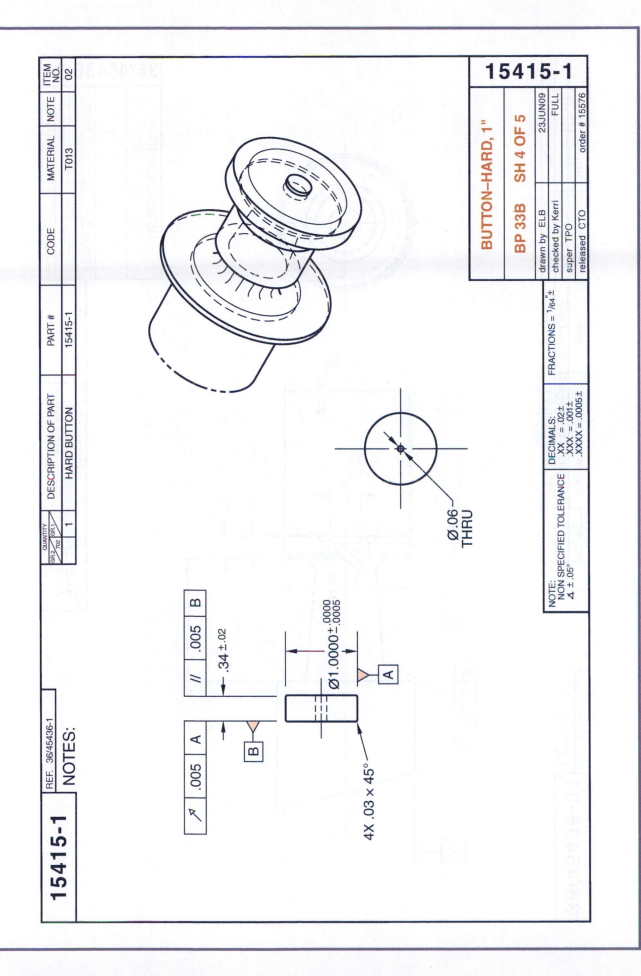

15415-1

BUTTON–HARD, 1"

BP 33B   SH 4 OF 5

drawn by ELB	23JUN09	
checked by Kerri	FULL	
super TPO		
released CTO	order # 15576	

FRACTIONS = ¹/₆₄ ±

DECIMALS:
.XX = .02±
.XXX = .001±
.XXXX = .0005±

NOTE:
NON SPECIFIED TOLERANCE
∢ ± .05°

Ø.06
THRU

ITEM NO.	02
NOTE	
MATERIAL	T013
CODE	
PART #	15415-1
DESCRIPTION OF PART	HARD BUTTON
QUANTITY GR 2 702 / GR 1	1

15415-1

REF. 36/45436-1

NOTES:

∥	.005	B

.34 ± .02

Ø1.0000 +.0000 −.0005

[A]

⌰	.005	A

[B]

4X .03 × 45°

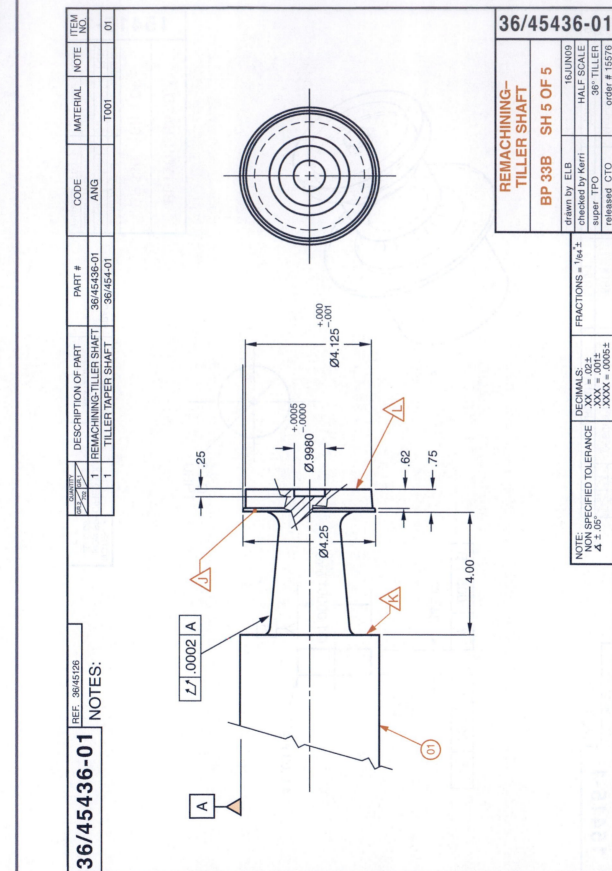

36/45436-01

REF. 36/45126 NOTES:

A

⊾ .0002 A

.25

Ø.9980 +.0005 −.0000

Ø4.125 +.000 −.001

Ø4.25

.62

.75

4.00

J

L

K

01

ITEM NO.	NOTE	MATERIAL	CODE	PART #	DESCRIPTION OF PART	QUANTITY	
						GR 2 702	GR 1
01		T001	ANG	36/45436-01	REMACHINING-TILLER SHAFT	1	
				36/454-01	TILLER TAPER SHAFT	1	

36/45436-01

REMACHINING– TILLER SHAFT

BP 33B   SH 5 OF 5

drawn by ELB	16JUN09
checked by Kerri	HALF SCALE
super TPO	36° TILLER
released CTO	order # 15576

NOTE:
NON SPECIFIED TOLERANCE
∡ ± .05°

FRACTIONS = 1/64"±

DECIMALS:
.XX = .02±
.XXX = .001±
.XXXX = .0005±

ASSIGNMENT B1—UNIT 33: TILLER TAPER
SHAFT ASSEMBLY (BP-33B)

Student's Name _____

Note: The following questions relate to the Assembly Drawing.

1. Print the name of the following parts:
   A, B, C, D.

   1. A _____
      B _____
      C _____
      D _____

2. Give the drawing number for the tiller taper shaft assembly.

   2. _____

3. How many detail and assembly drawings make up the tiller taper shaft working drawing?

   3. _____

4. How many parts are needed to make a tiller taper shaft assembly?

   4. _____

5. How many of these parts are commercial or purchased parts?

   5. _____

Note: The following questions relate to the Bearing.

1. Tell why the section lines of (04) go in two (2) different directions.

   1. _____

2. Give the minimum taper angle of the bearing.

   2. _____

3. What is the part number of the bearing?

   3. _____

4. How many pieces make up the bearing?

   4. _____

5. What is the basic height and thickness of the bearing?

   5. _____

Note: The following questions relate to the Ring.

1. What is the basic thickness of the ring?

   1. _____

2. What is the basic OD of the ring?

   2. _____

3. What is the minimum hole size of the ring when using the upper-limit of the shaft and the lower-limit of the ring?

   3. _____

4. Explain a possible meaning of "1.0000 Hold".

   4. _____

5. Explain why the 2.00 REF is given.

   5. _____

ASSIGNMENT B2—UNIT 33:

Student's Name _____

Note: The following questions relate to the GEAR.

1. What surface finish sizes are given on the cast gear?

   1. _____

2. What is the smoothest surface of the gear?

   2. _____

3. What do the hidden lines at Ⓔ indicate?

   3. _____

4. Give the part number of the gear.

   4. _____

5. What manufacturing process is used to make the gear?

   5. _____

6. Using a reference handbook, give the following gear terms: Ⓖ, Ⓗ, Ⓘ.

   6. Ⓖ _____

   Ⓗ _____

   Ⓘ _____

7. How many teeth does the gear have?

   7. _____

8. Determine the basic clearance between the tiller shaft and the .13 gear slot diameter.

   8. _____

Note: The following questions relate to the BUTTON.

1. Research, then explain why part 01 will contract, allowing the large part to fit into the smaller part.

   1. _____

   _____

   _____

2. Calculate the allowance between the tiller taper shaft and the button.

   2. _____

3. Explain what each item in the feature control frames for the button mean:

   3. _____

   _____

   _____

   A) | ⌐ | .005 | B |

   B) | // | .005 | A |

Note: The following questions relate to the SHAFT.

1. Determine the lower-limit of the thickness of △J.

   1. _____

2. Explain what each item in the feature control frame of the tiller shaft means. (See Unit 29.)

   2. _____

   _____

3. Determine the upper-limit distance between surfaces △K and △L.

   3. _____

15415-1 | T013 | 02

Technical Sketching

Sketching Lines and Basic Forms

Sketching Horizontal, Vertical, and Slant Lines

THE VALUE OF SKETCHING

Technical sketches are used in all phases of manufacturing and constructing—from concept development to product completion to marketing and sales to after-market repair and maintenance.

They are an essential aid in the visualization process because they can be made quickly anywhere in the world, needing only a sketching instrument and a material to sketch upon. Sketches, at times, may be used in place of instrument drawings. In Part 2 of this text, all the basics of technical sketching will be covered and experience in this will be given through completing specially devised sketching assignments.

SKETCHING LINES FREEHAND

The development of correct skills in sketching lines freehand is more essential for the beginner than speed. After the basic principles of sketching are learned and skill is acquired in making neat and accurate sketches, then stress should be placed on speed.

Although shop sketches are made on the job and special pencils are not required, a medium-soft to medium (HB, No. 2) lead pencil with a cone-shaped point will produce good results. For fine lines, use a fairly sharp point; for heavier lines, round the point more, Figure 34–1. A sharp point may be kept longer by rotating the pencil. When available, adjustable 0.5, 0.7, and 0.9 mm pencils with HB or F lead and soft white click erasers can produce correct line thickness and darkness and excellent quality sketches.

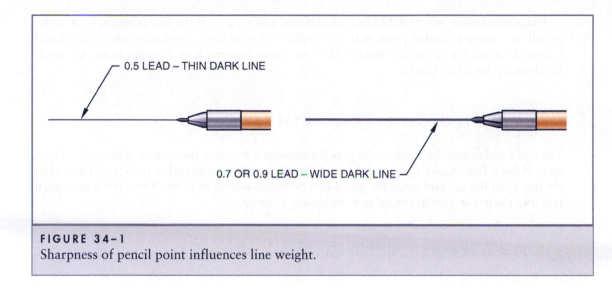

FIGURE 34-1
Sharpness of pencil point influences line weight.

SKETCHING HORIZONTAL LINES

The same principles of drafting that apply to the making of a mechanical drawing and the interpretation of blueprints are used for making sketches.

In sketching, the pencil is held 3/4" to 1" from the point so the lines to be drawn may be seen easily and there is a free and easy movement of the pencil. A free arm movement makes it possible to sketch smooth, neat lines as compared with the rough and inaccurate lines produced by a finger and wrist movement.

In planning the first horizontal line on a sketch, place two points on the paper to mark the beginning and end of the line, Figure 34–2A. Place the pencil point on the first dot. Then, with a free arm movement and with the eyes focused on the right point, draw the line from left to right, Figure 34–2B.

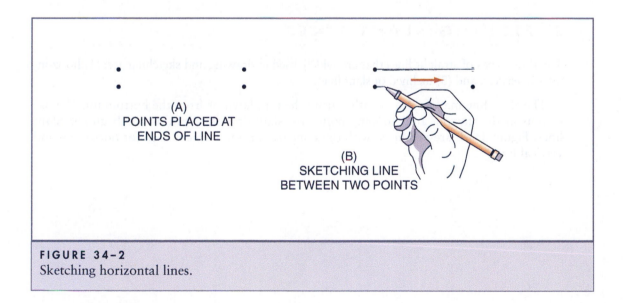

FIGURE 34-2
Sketching horizontal lines.

Examine the line for straightness, smoothness, and weight. If the line is too light, a softer pencil or a more rounded point may be needed. On long lines, extra dots are often placed between the start and finish points of the line. These intermediate dots are used as a guide for drawing long straight lines.

SKETCHING VERTICAL LINES

The same techniques for holding the pencil and using a free arm movement apply in sketching vertical lines. Dots again may be used to indicate the beginning and end of the vertical line. Start the line from the top and move the pencil downward, as shown in Figure 34–3. For long lines, if possible, the paper may be turned to a convenient position.

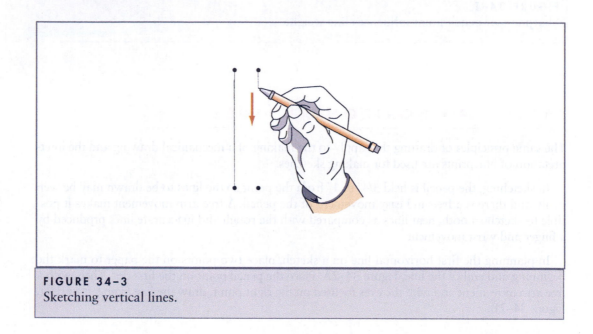

FIGURE 34–3
Sketching vertical lines.

SKETCHING SLANT LINES

The three types of straight lines that are widely used in drawing and sketching are: (1) horizontal, (2) vertical, and (3) inclined or slant lines.

The slant line may be drawn either from the top down or from the bottom up. The use of dots at the starting and stopping points for slant lines is helpful to the beginner. Slant lines, Figure 34–4, are produced with the same free arm movement used for horizontal and vertical lines.

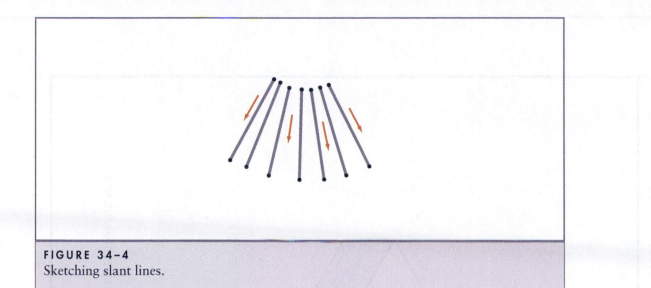

FIGURE 34–4
Sketching slant lines.

USING GRAPH PAPER

Many industries recommend the use of graph paper for making sketches. Graph paper is marked to show how many squares are included in each inch. The combination of light ruled lines at a fixed number per inch makes it possible to draw neat sketches that are fairly accurate in size. Throughout this text, squared or isometric grids are included for each sketching assignment. Other ruled grid papers are available with diametric, oblique, or perspective lines.

PUNCH PLATE **BP-34A**

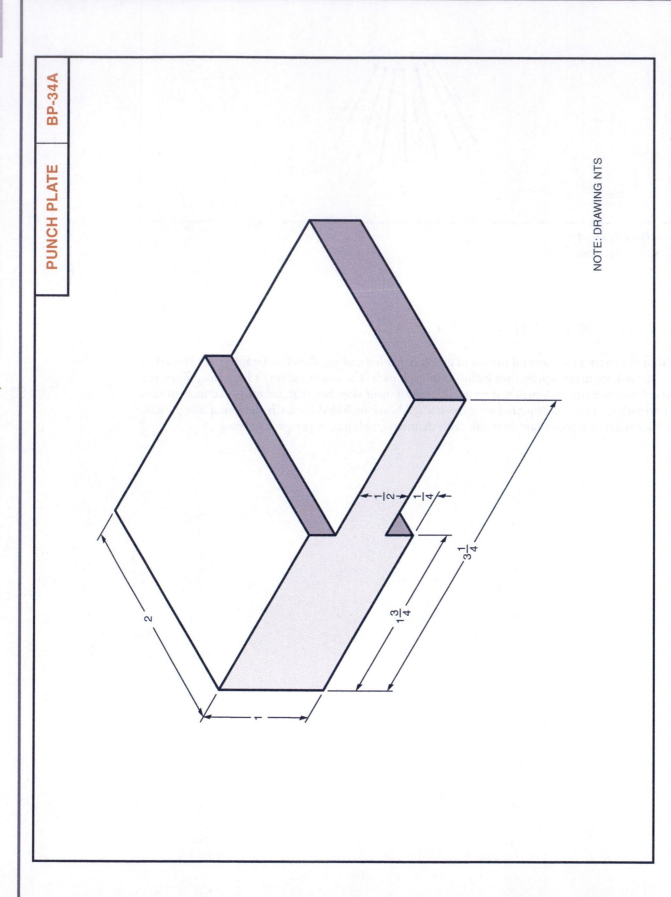

NOTE: DRAWING NTS

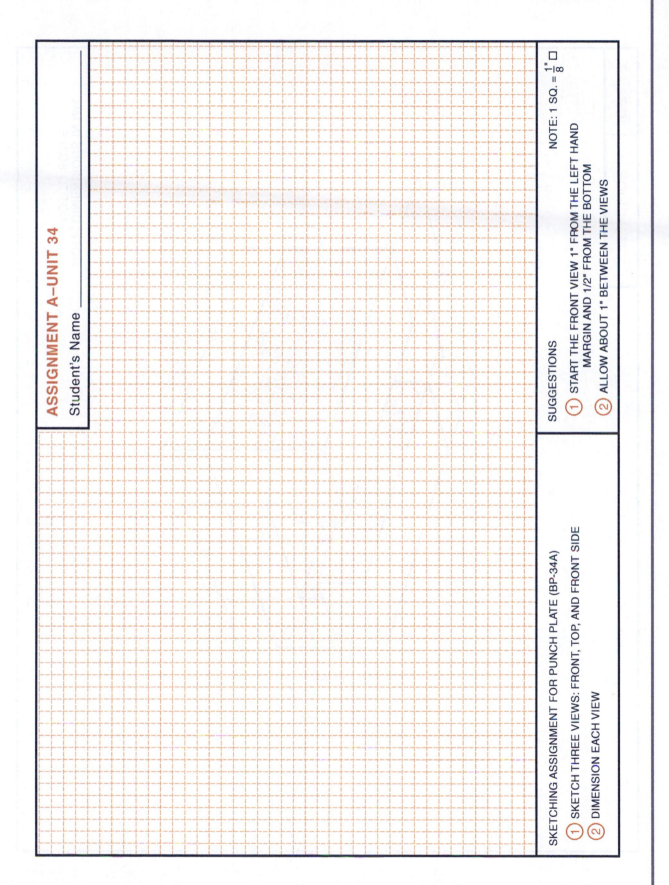

ASSIGNMENT A–UNIT 34

Student's Name

NOTE: 1 SQ. = $\frac{1''}{8}$ □

SKETCHING ASSIGNMENT FOR PUNCH PLATE (BP-34A)

① SKETCH THREE VIEWS: FRONT, TOP, AND FRONT SIDE

② DIMENSION EACH VIEW

SUGGESTIONS

① START THE FRONT VIEW 1" FROM THE LEFT HAND MARGIN AND 1/2" FROM THE BOTTOM

② ALLOW ABOUT 1" BETWEEN THE VIEWS

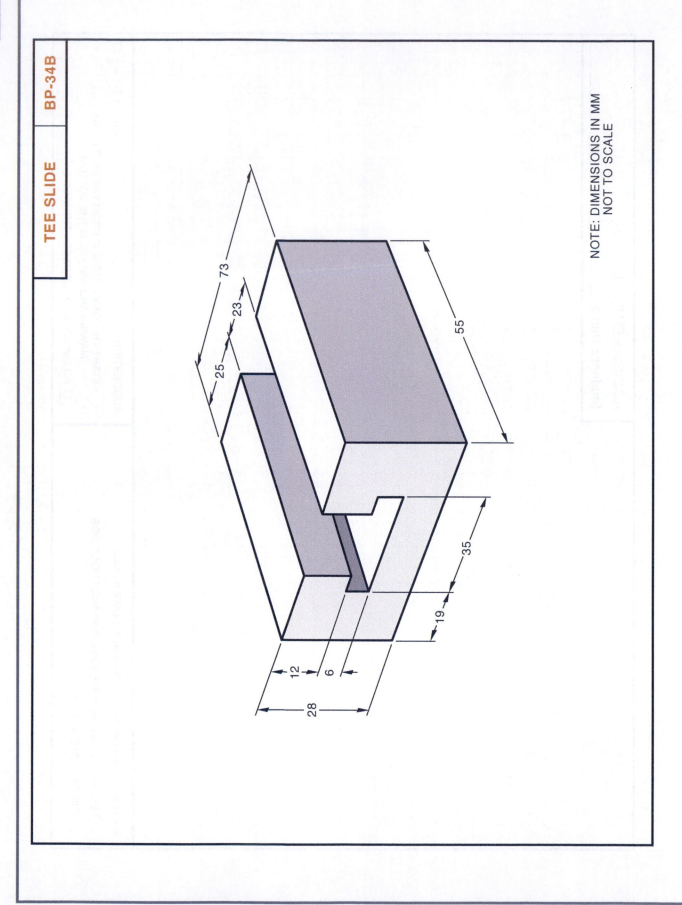

TEE SLIDE | **BP-34B**

NOTE: DIMENSIONS IN MM
NOT TO SCALE

ASSIGNMENT B—UNIT 34

Student's Name _____

NOTE: 1 SQ. = 10MM □
GRID N.T.S.

SUGGESTIONS

① START FRONT VIEW 25MM FROM
LEFT-HAND MARGIN AND 20MM FROM BOTTOM

② ALLOW ABOUT 25MM BETWEEN THE VIEWS

SKETCHING ASSIGNMENT FOR TEE SLIDE (BP-34B)

① MAKE FREEHAND SKETCHES OF THE FRONT, TOP, AND
RIGHT-SIDE VIEWS

② DIMENSION THE VIEWS

UNIT 34 | SKETCHING HORIZONTAL, VERTICAL, AND SLANT LINES

35

Sketching Curved Lines and Circles

ARCS CONNECTING STRAIGHT LINES

Often, when describing a part, it is necessary to draw both straight and curved lines. When part of a circle is shown, the curved line is usually called an **arc.**

When an arc must be drawn so that it connects two straight lines (in other words, is tangent to them), it may be sketched easily if five basic steps are followed, Figure 35–1.

STEP 1 Extend the straight lines so they intersect.

STEP 2 Step off the same distance from the intersection (center) of both lines.

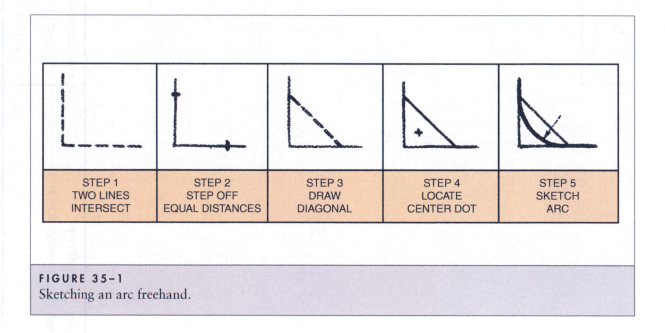

STEP 1	STEP 2	STEP 3	STEP 4	STEP 5
TWO LINES INTERSECT	STEP OFF EQUAL DISTANCES	DRAW DIAGONAL	LOCATE CENTER DOT	SKETCH ARC

FIGURE 35–1
Sketching an arc freehand.

STEP 3 Draw the diagonal line through these two points to form a triangle.

STEP 4 Place a dot in the center of the triangle.

STEP 5 Start at one of the lines and sketch an arc that runs through the dot and ends on the other line. Darken the arc and erase all unnecessary lines.

ARCS CONNECTING STRAIGHT AND CURVED LINES

Two other line combinations using straight lines and arcs are also very common. The first is the case of an arc connecting a straight line with another arc; the second is where an arc connects two other arcs. The same five basic steps are used as illustrated in Figure 35–2.

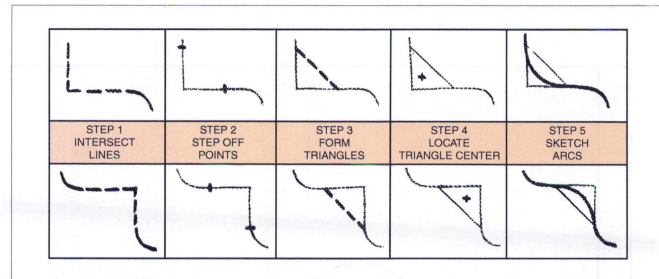

FIGURE 35-2
Sketching arcs in combination with other arcs.

SKETCHING CIRCLES

Shop sketches also require the drawing of circles or parts of a circle. While there are many ways to draw a circle freehand, a well-formed circle can be sketched by following five simple steps that are described and illustrated in Figure 35–3.

STEP 1 Lay out the vertical and horizontal center lines in the correct location. Measure off half the diameter of the circle on each side of the two center lines.

STEP 2 Lightly draw two vertical and two horizontal lines passing through the points marked off on the center lines. These lines, properly drawn, will be parallel and will form a square.

STEP 3 Draw diagonal lines from each corner of the square to form four triangles.

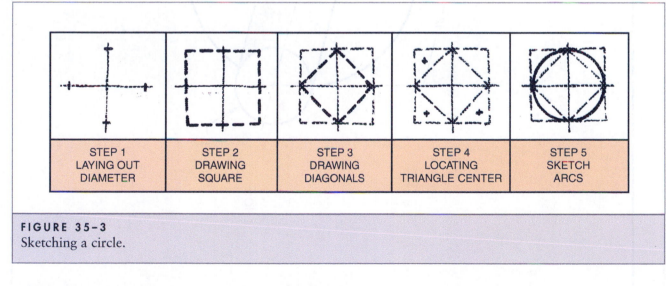

FIGURE 35-3
Sketching a circle.

STEP 4 Place a dot in the center of each triangle through which an arc is to pass.

STEP 5 Sketch the arc for one quarter of the circle. Start at one center line and draw the arc through the dot to the next center line. Continue until the circle is complete. Darken the circle and erase all the guide lines to simplify the reading of the sketch.

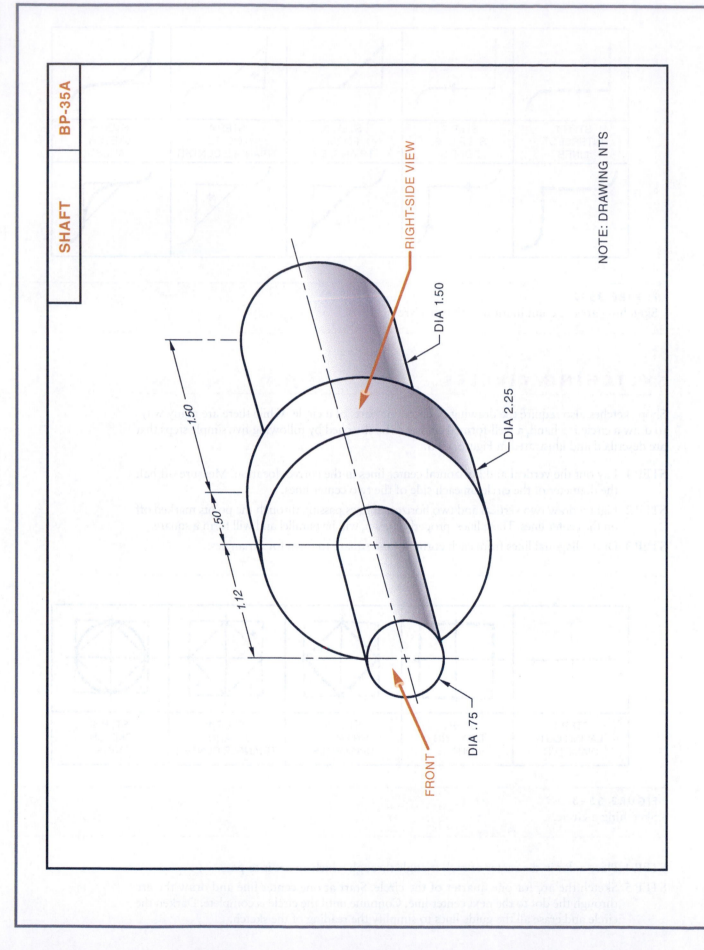

SHAFT   BP-35A

RIGHT-SIDE VIEW

DIA 1.50

DIA 2.25

1.50

.50

1.12

DIA .75

FRONT

NOTE: DRAWING NTS

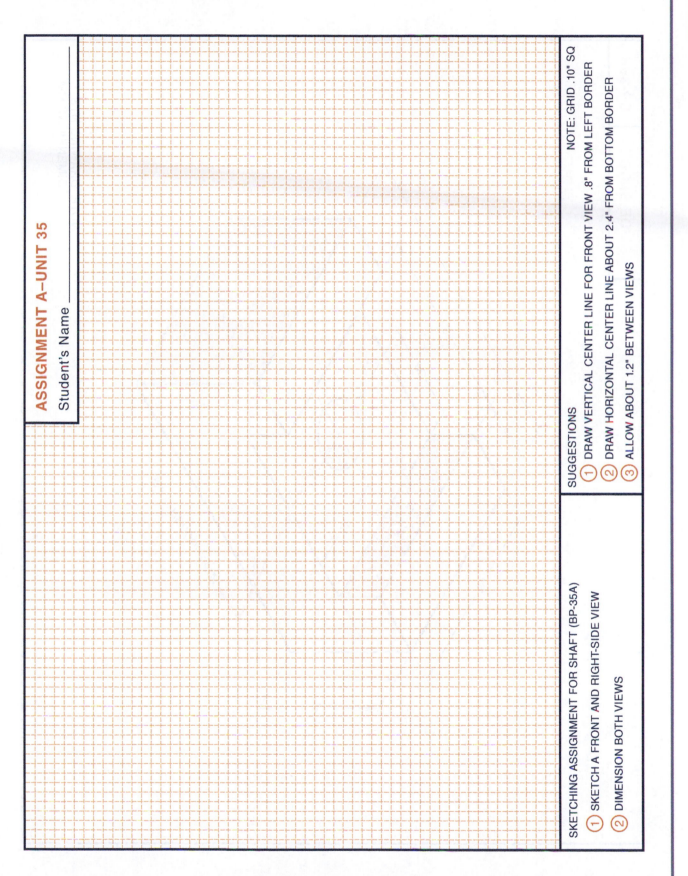

ASSIGNMENT A–UNIT 35

Student's Name _____

SKETCHING ASSIGNMENT FOR SHAFT (BP-35A)

(1) SKETCH A FRONT AND RIGHT-SIDE VIEW

(2) DIMENSION BOTH VIEWS

NOTE: GRID .10" SQ

SUGGESTIONS

(1) DRAW VERTICAL CENTER LINE FOR FRONT VIEW .8" FROM LEFT BORDER

(2) DRAW HORIZONTAL CENTER LINE ABOUT 2.4" FROM BOTTOM BORDER

(3) ALLOW ABOUT 1.2" BETWEEN VIEWS

UNIT 35 | SKETCHING CURVED LINES AND CIRCLES

SLIDE BLOCK | BP-35B

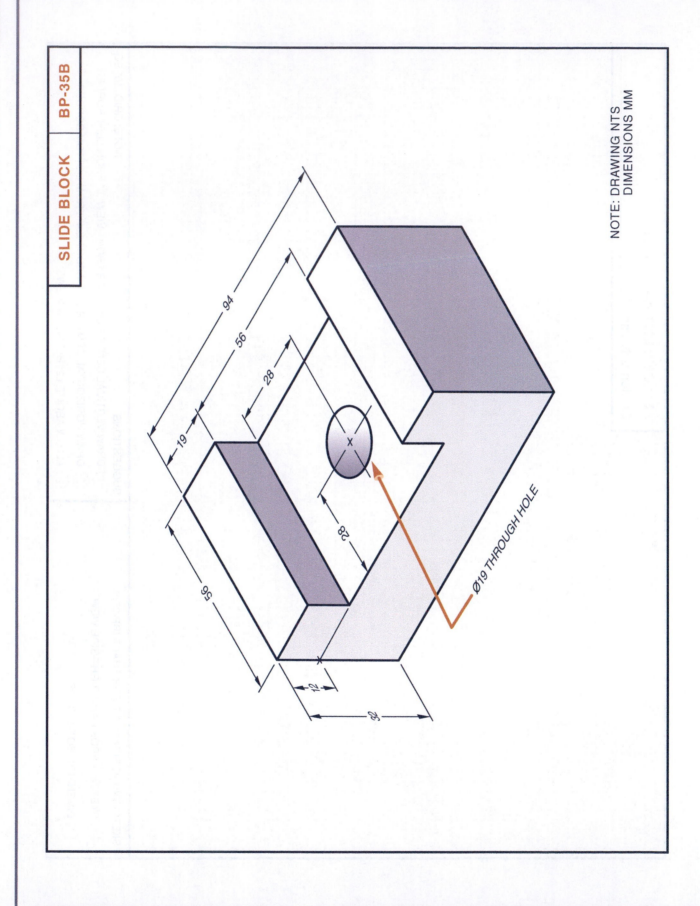

94
56
28
19
56
28
12
32

Ø19 THROUGH HOLE

X

NOTE: DRAWING NTS
DIMENSIONS MM

ASSIGNMENT B—UNIT 35

Student's Name

SKETCHING ASSIGNMENT FOR SLIDE BLOCK (BP-35B)

1. SKETCH THREE VIEWS: FRONT, TOP AND RIGHT-SIDE
2. DIMENSION EACH VIEW OF THE SKETCH

NOTE: GRID 10MM SQ

SUGGESTIONS

1. START THE FRONT VIEW ABOUT 15MM FROM THE LEFT-HAND MARGIN AND ABOUT 12MM FROM THE BOTTOM
2. ALLOW ABOUT 25MM BETWEEN THE VIEWS

FLANGE

BP-35C

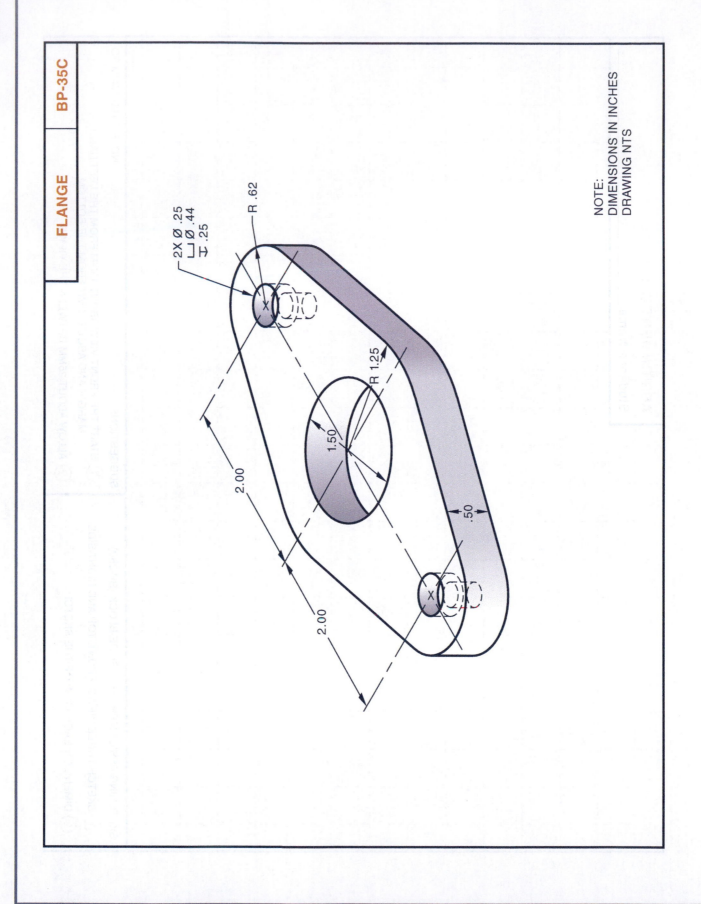

2X Ø .25
⌴ Ø .44
⊤ .25

R .62

R 1.25

1.50

2.00

2.00

.50

NOTE:
DIMENSIONS IN INCHES
DRAWING NTS

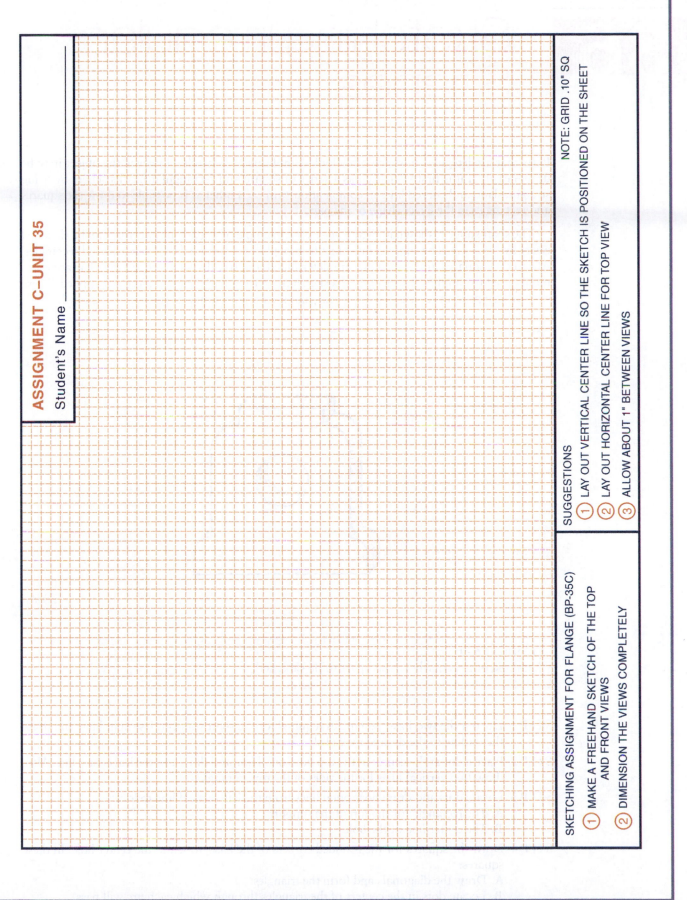

ASSIGNMENT C–UNIT 35

Student's Name

NOTE: GRID .10" SQ

SUGGESTIONS

① LAY OUT VERTICAL CENTER LINE SO THE SKETCH IS POSITIONED ON THE SHEET

② LAY OUT HORIZONTAL CENTER LINE FOR TOP VIEW

③ ALLOW ABOUT 1" BETWEEN VIEWS

SKETCHING ASSIGNMENT FOR FLANGE (BP-35C)

① MAKE A FREEHAND SKETCH OF THE TOP AND FRONT VIEWS

② DIMENSION THE VIEWS COMPLETELY

UNIT 35 | SKETCHING CURVED LINES AND CIRCLES

36

Sketching Irregular Shapes

Parts that are irregular in shape often look complicated to sketch. However, the object may be drawn easily if it is first visualized as a series of square and rectangular blocks. Then, by using straight, slant, and curved lines in combination with each other, it is possible to draw the squares and rectangles in the exact shape of the part.

The shaft support shown in Figure 36–1 is an example of a machine part that can be sketched easily by blocking. The support must first be thought of in terms of basic squares and rectangles. After these squares and rectangles are determined, the step-by-step procedures that are commonly used to simplify the making of a sketch of an irregularly shaped piece are applied. These step-by-step procedures are described next, and their application is shown in Figure 36–2.

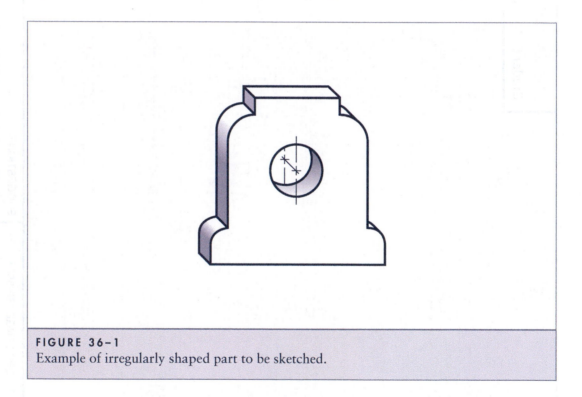

FIGURE 36–1
Example of irregularly shaped part to be sketched.

STEP 1 Lay out the two basic rectangles required for the shaft support. Use light lines and draw the outline to the desired overall sketch size.

STEP 2 Place a dot at the center of the top horizontal line and draw the vertical center line. Also, draw two vertical lines for the sides of the top rectangle.

STEP 3 Locate dots and draw horizontal lines from the left vertical line to locate:
A. The rectangle for the top of the support
B. The center line of the hole

STEP 4 Draw the squares for the rounded corners of the base, top, and the center hole. In these squares:
A. Draw the diagonals and form the triangles.
B. Locate dots in the centers of the triangles through which each arc will pass.

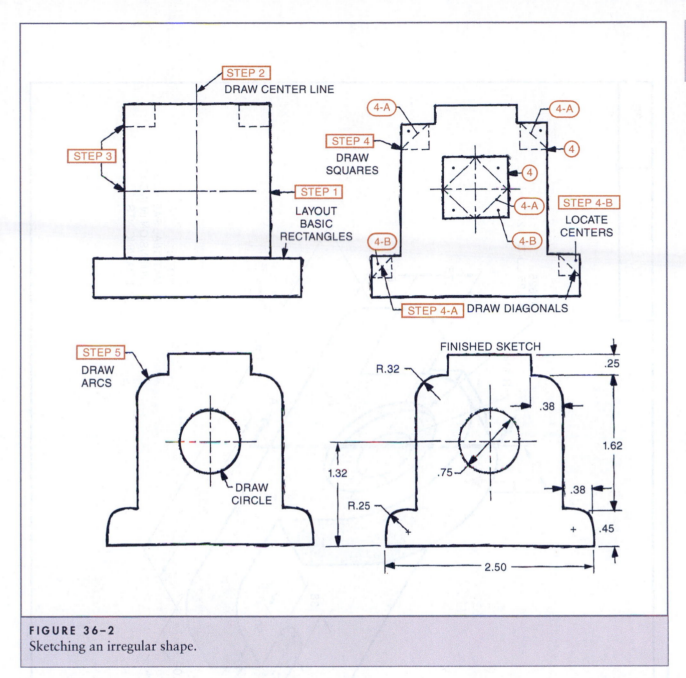

FIGURE 36-2
Sketching an irregular shape.

STEP 5 Draw the arcs and circle. Start at one diagonal and draw the arc through the dot to the next diagonal. Continue until the circle is completed.

STEP 6 Draw the side view if necessary.

STEP 7 Darken all object lines and dimension. Erase those lines used in construction that either do not simplify the sketch or are not required to interpret the sketch quickly and accurately.

The techniques described in this unit have been found by tested experience to be essential in training the beginner to sketch accurately.

As skill is developed in drawing straight, slant, and curved lines freehand in combination with each other, some of the steps in sketching irregularly shaped parts may be omitted and the process shortened.

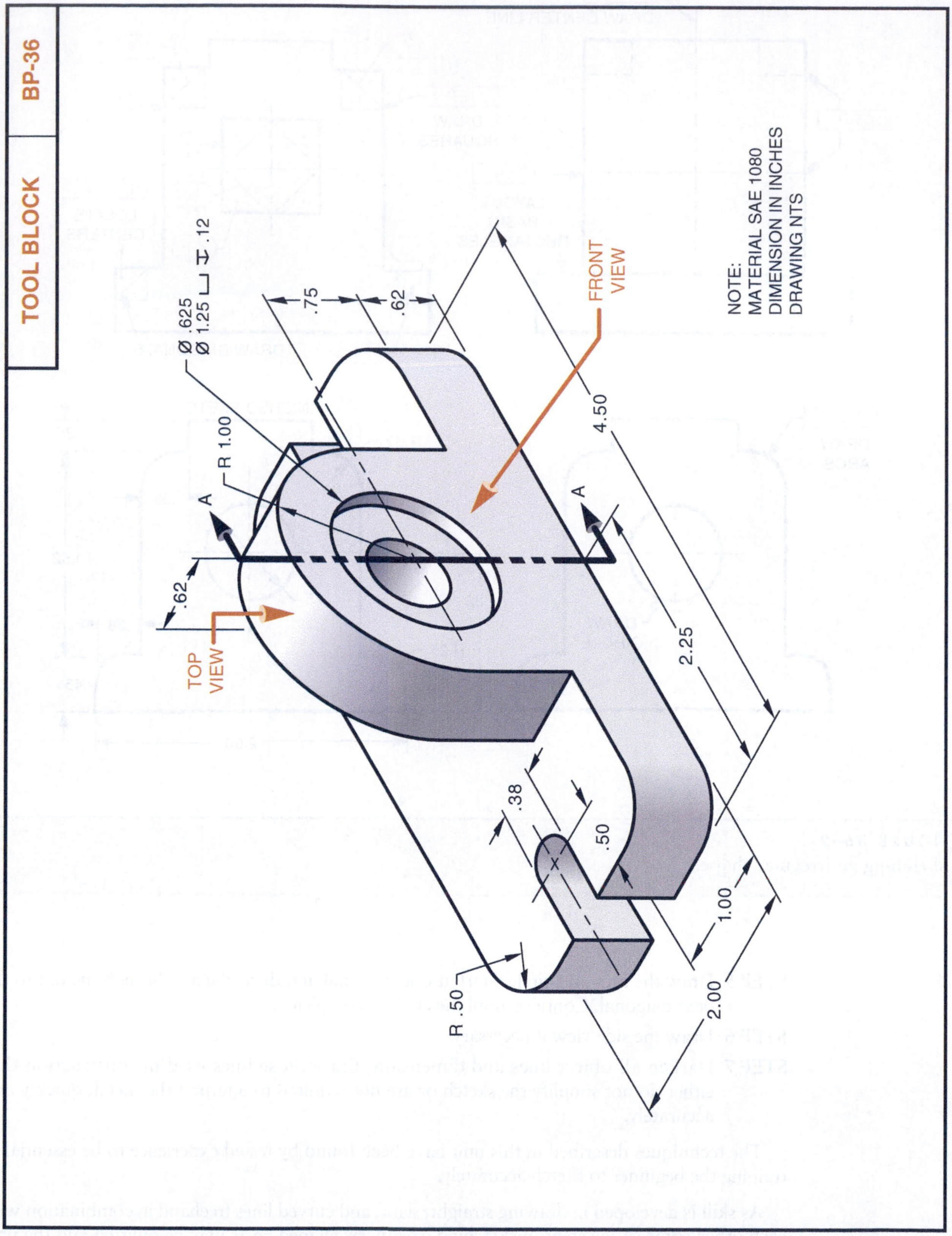
TOOL BLOCK
BP-36
Ø .625
Ø 1.25 ⌴ ↧ .12
.75
.62
R 1.00
4.50
FRONT VIEW
NOTE:
MATERIAL SAE 1080
DIMENSION IN INCHES
DRAWING NTS
A
A
.62
TOP VIEW
2.25
.38
.50
1.00
R .50
2.00

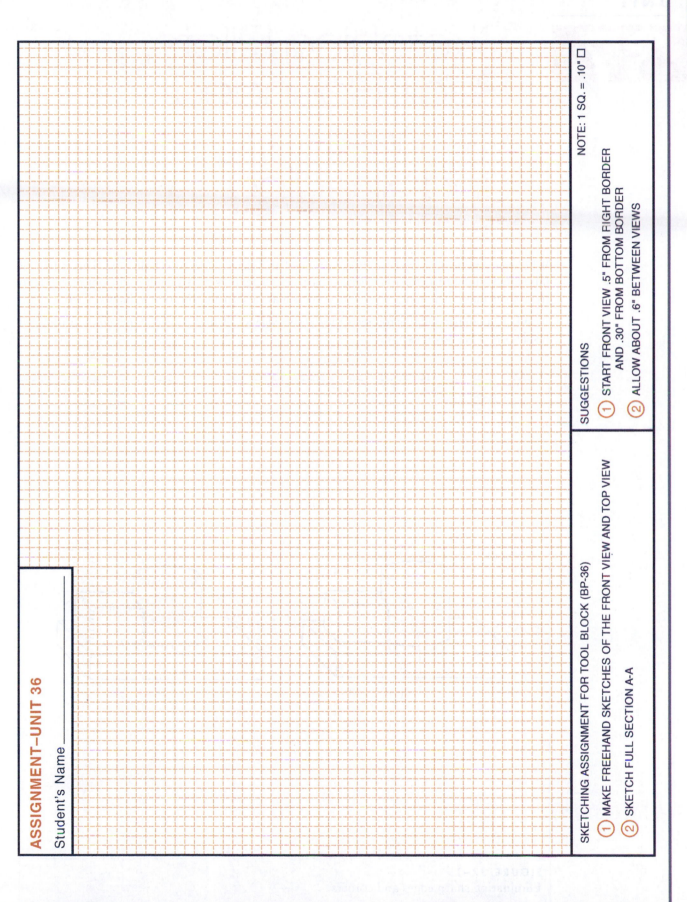

ASSIGNMENT–UNIT 36

Student's Name _____

SKETCHING ASSIGNMENT FOR TOOL BLOCK (BP-36)

① MAKE FREEHAND SKETCHES OF THE FRONT VIEW AND TOP VIEW

② SKETCH FULL SECTION A-A

SUGGESTIONS

NOTE: 1 SQ. = .10" ☐

① START FRONT VIEW .5" FROM FIGHT BORDER
   AND .30" FROM BOTTOM BORDER

② ALLOW ABOUT .6" BETWEEN VIEWS

UNIT 36 | SKETCHING IRREGULAR SHAPES

37

Sketching Fillets, Radii, Rounded Corners, Edges, and Shading

Wherever practical, sharp corners are rounded and sharp edges are eliminated. This is true of fabricated and hardened parts where a sharp corner reduces the strength of the object. It is also true of castings where sharp edges and corners may be difficult to produce, Figure 37–1.

Where an outside edge is rounded, the convex edge is called a round edge or radius. A rounded inside corner is known as a fillet.

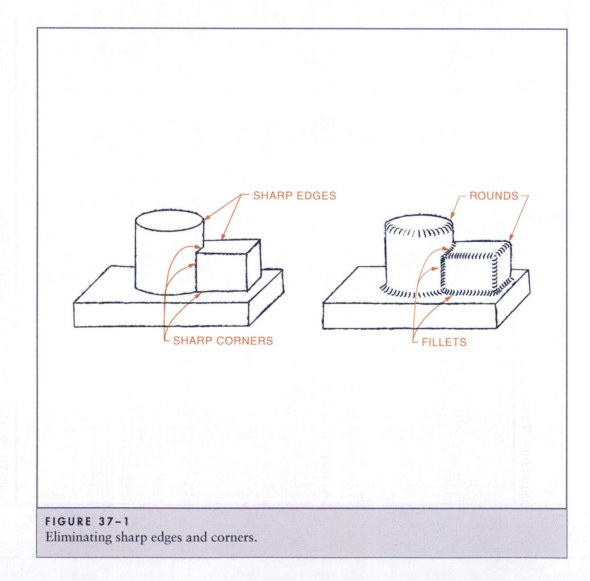

FIGURE 37–1
Eliminating sharp edges and corners.

SKETCHING A FILLET OR RADIUS

Regardless of whether a fillet or radius is required, the steps for sketching each one are identical. The step-by-step procedure is illustrated in Figure 37–2.

STEP 1 Sketch the lines that represent the edge or corner.

STEP 2 Draw the arc to the required radius.

STEP 3 Draw two lines parallel to the edge line from the two points where the arc touches the straight lines.

STEP 4 Sketch a series of curved lines across the work with the same arc as the object line. These curved lines start at one of the parallel lines and terminate at the other end.

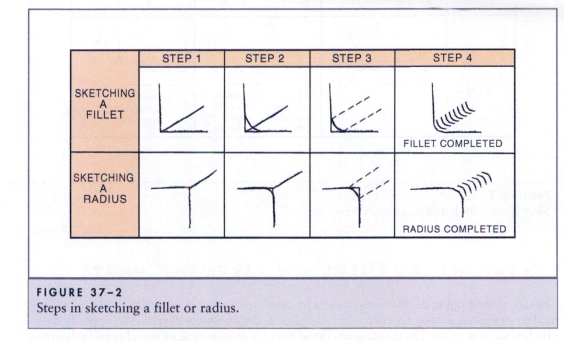

FIGURE 37–2
Steps in sketching a fillet or radius.

SKETCHING RADII AND FILLETS

The method of sketching corners at which radii or fillets come together is shown in the three steps in Figure 37–3.

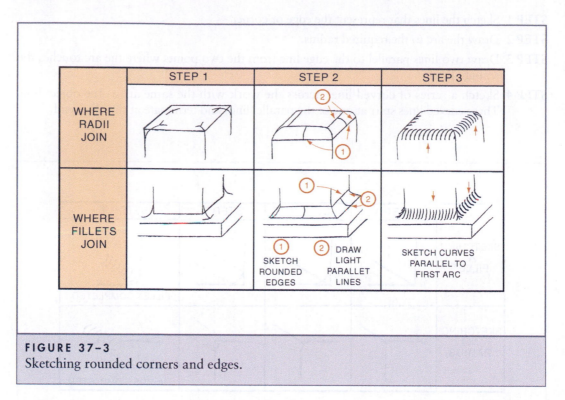

FIGURE 37-3
Sketching rounded corners and edges.

SKETCHING CORNERS ON CIRCULAR PARTS

The direction of radius of fillet lines changes on circular objects at a center line. The curved lines tend to straighten as they approach the center line and then slowly curve in the opposite direction beyond that point. The direction of curved lines for outside radii of round parts is shown in Figure 37–4A. The curved lines for the intersecting corners are shown in Figure 37–4B.

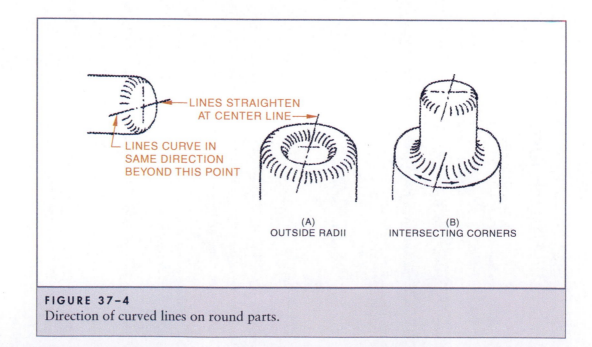

FIGURE 37-4
Direction of curved lines on round parts.

BASIC SHADING/RENDERING TECHNIQUES

Shading (sometimes referred to as rendering) is often used on drawings that have hard-to-read part features. Where required, as a general rule, the draftsperson uses the least amount of shading possible. Shading may be produced freehand, generated by computer, or created by cutting out cross-hatch, dot, or other shade line patterns from an acetate sheet and pressing the shade material on a drawing.

Finely spaced, thin, straight lines and/or concentric finely spaced curved lines or narrow, solid bands are used for shading. Figure 37–5 shows how shade lines and solid shade bands clearly identify the end radius, should fillet, cylindrical body, and keyseat features of the part. For more shading techniques, see Unit 43.

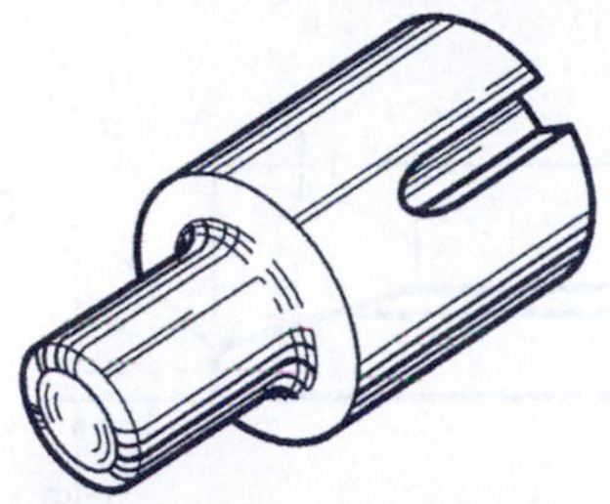

FIGURE 37–5
Shading with straight and curved lines and narrow, solid bands.

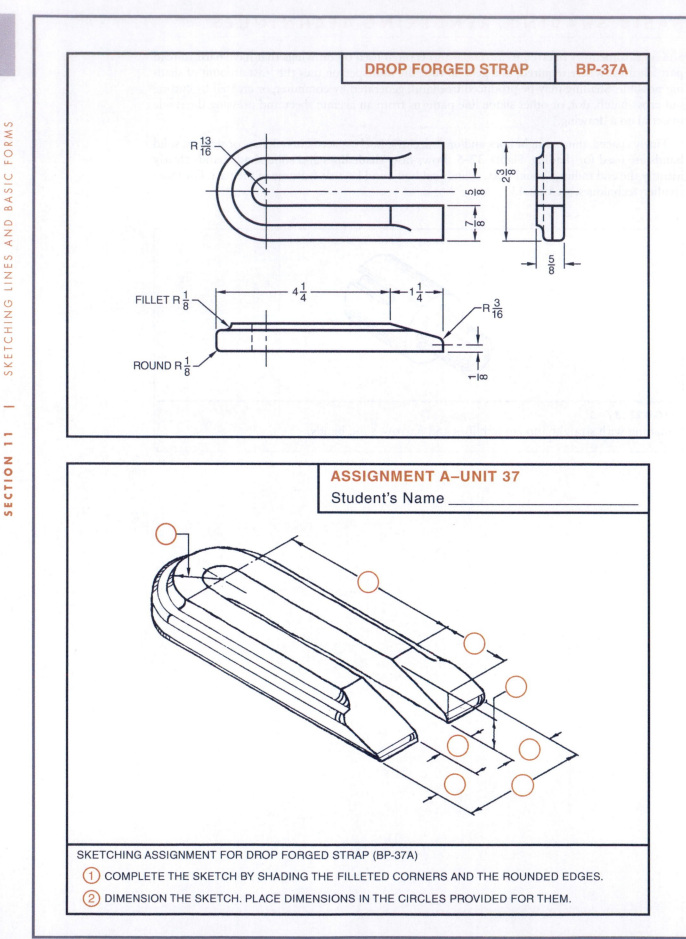

| DROP FORGED STRAP | BP-37A |

R$\frac{13}{16}$

2$\frac{3}{8}$

5$\frac{5}{8}$

7$\frac{7}{8}$

$\frac{5}{8}$

FILLET R$\frac{1}{8}$

4$\frac{1}{4}$

1$\frac{1}{4}$

R$\frac{3}{16}$

ROUND R$\frac{1}{8}$

1$\frac{1}{8}$

ASSIGNMENT A–UNIT 37
Student's Name _____

SKETCHING ASSIGNMENT FOR DROP FORGED STRAP (BP-37A)

① COMPLETE THE SKETCH BY SHADING THE FILLETED CORNERS AND THE ROUNDED EDGES.

② DIMENSION THE SKETCH. PLACE DIMENSIONS IN THE CIRCLES PROVIDED FOR THEM.

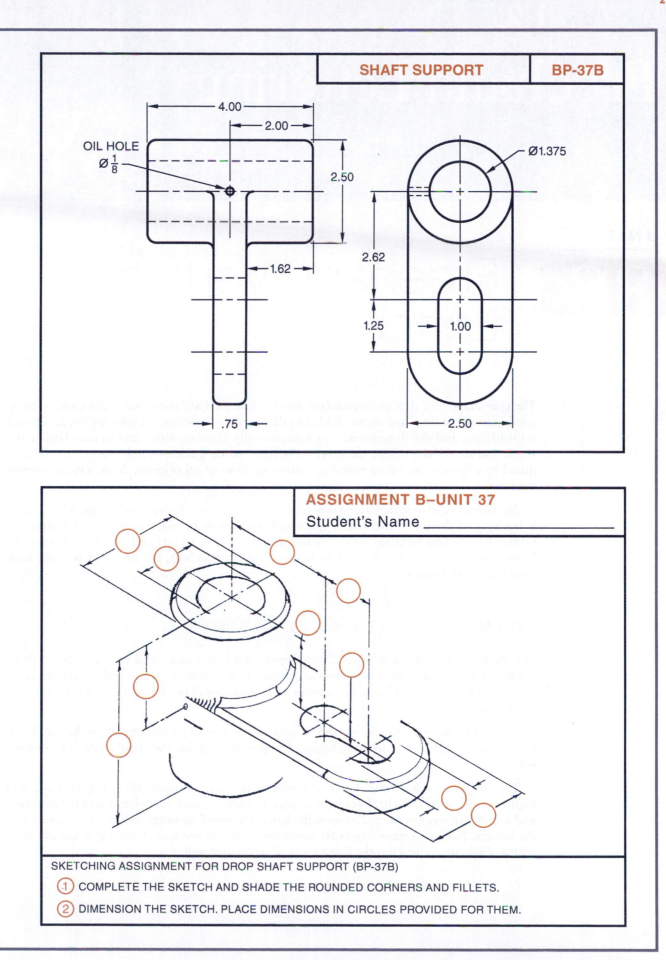

SHAFT SUPPORT | BP-37B

OIL HOLE
Ø 1/8

4.00

2.00

2.50

1.62

.75

Ø1.375

2.62

1.25

1.00

2.50

ASSIGNMENT B–UNIT 37

Student's Name _____

SKETCHING ASSIGNMENT FOR DROP SHAFT SUPPORT (BP-37B)

① COMPLETE THE SKETCH AND SHADE THE ROUNDED CORNERS AND FILLETS.

② DIMENSION THE SKETCH. PLACE DIMENSIONS IN CIRCLES PROVIDED FOR THEM.

Freehand Lettering

38

Freehand Vertical Lettering

The true shape of a part or mechanism may be described accurately on a drawing by using combinations of lines and views. Added to these are the lettering, which supplies additional information, and the dimensions. For exceptionally accurate work and to standardize the shape and size of the letters, the lettering is done with a guide or the lettering may be produced by computer and then added to a drawing. In most other cases, the letters are formed freehand.

An almost square style letter known as Gothic lettering is very widely used because it is legible and the individual letters are simple enough to be made quickly and accurately. Gothic letters may be either vertical (straight) or inclined (slant) and upper- or lowercase. Lower-case letters are rarely used on manufacturing drawings, but are used at times with construction drawings.

FORMING UPPER-CASE LETTERS

The shape of each vertical letter (both upper- and lower-case) and each number will be discussed in this unit. All these letters and numbers are formed by combining vertical, horizontal, slant, and curved lines. The upper-case letters should be started first, as they are easiest to make.

Very light guide lines should be used to keep the letters straight and of uniform height. A soft pencil is recommended for lettering because it is possible to guide the pencil easily to form good letters.

The shape of each vertical letter and number is given in Figure 38–1, Figure 38–2, and Figure 38–3. The fine (light) lines with arrows that appear with every letter give the direction and number of strokes needed to form the letter. The small numbers indicate the sequence of the strokes. Note that most letters are narrower than they are long. Lettering is usually done with a single stroke to keep the line weight of each letter uniform.

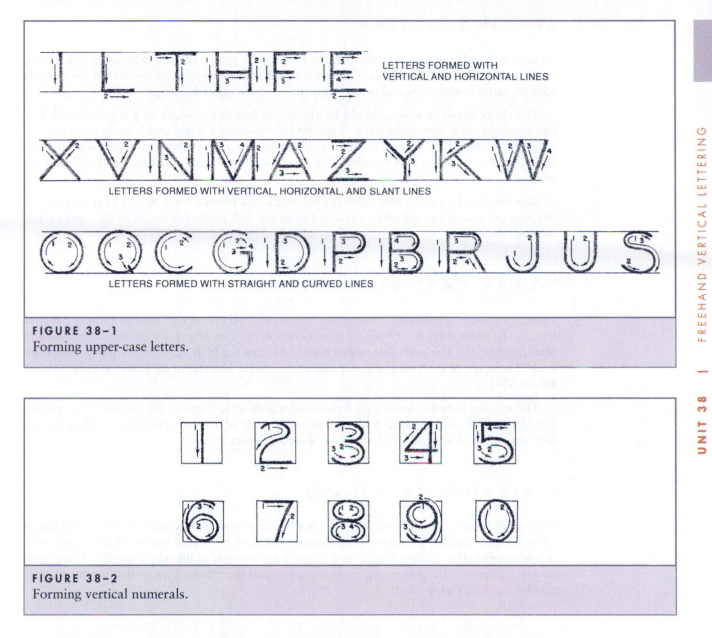

FIGURE 38-1
Forming upper-case letters.

FIGURE 38-2
Forming vertical numerals.

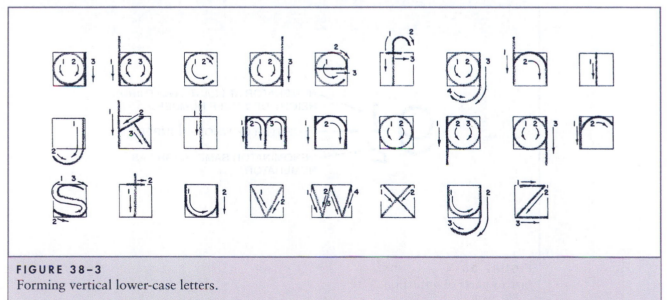

FIGURE 38-3
Forming vertical lower-case letters.

SPACING LETTERS

To achieve good lettering, attention must be given to the proper spacing between letters, words, and lines. Words and lines that are either condensed and run together or spread out are difficult to read, cause inaccuracies, and detract from an otherwise good drawing.

The space between letters should be about one-fourth the width of a regular letter. For example, the slant line of the letter **A** should be one-quarter letter width away from the top line of the letter **T.** Judgment must be used in the amount of white space left between letters so that it is as equal as possible. The letters will then look in balance and will be easy to read.

Between words, a space two-thirds the full width of a normal letter should be used. Lines of lettering are easiest to read when a space of from one-half to the full height of the letters is left between the lines.

LOWER-CASE VERTICAL LETTERS

Lower-case vertical letters are formed in a manner different from that used to form upper-case letters. The main portion, or body, of most Gothic lower-case letters is approximately the same width and height. The parts that extend above or below the body are one-half the height of the body. The shape of each vertical lower-case letter and the forming of the letters are indicated in Figure 38–3.

The spacing between lower-case letters and words is the same as for capitals. The spacing between lines should be equal to the height of the body of a lower-case letter to allow for the lines extending above and below the body of certain letters.

LETTERING FRACTIONS

Because fractions are important, they must not be subordinate to any of the lettering. The height of each number in the numerator and denominator must be at least two-thirds the height of a whole number. The dividing line of the fraction is in the center of the whole number. There must be a space between the numerator, the dividing line, and the denominator so that neither number touches the line, Figure 38–4.

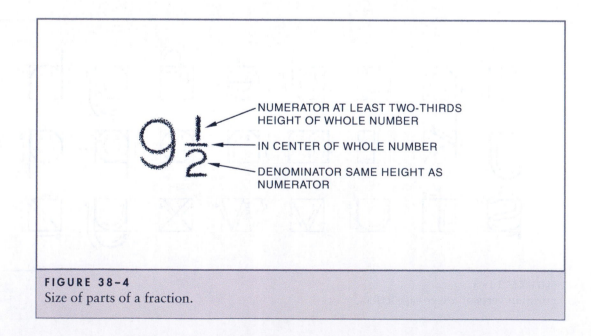

FIGURE 38–4
Size of parts of a fraction.

Lettering for Microfilm and Photocopies

Lettering used on drawings that are to be microfilmed or greatly reduced require special attention. The National Micrographics Association has developed an adaptation of the single-stroke Gothic characters for general use and increased legibility in drawing reproduction. The letters are known as microfont. Figure 38–5 shows only the vertical style. Some companies that use microfilming, however, continue to use standard letters, large and dense enough to clearly reproduce.

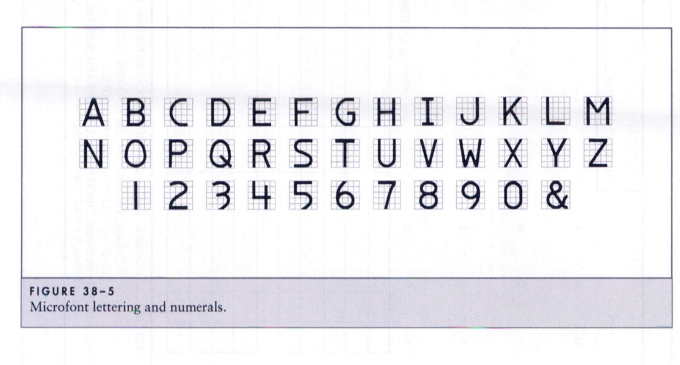

FIGURE 38–5
Microfont lettering and numerals.

ASSIGNMENT–UNIT 38

Student's Name _____

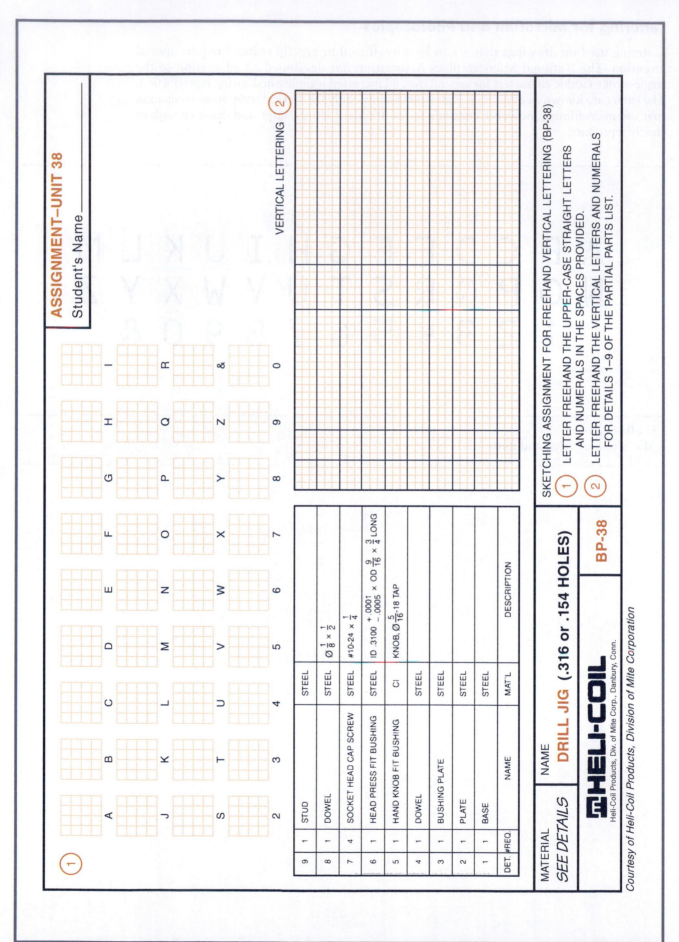

A | **B** | **C** | **D** | **E** | **F** | **G** | **H** | **I**

J | **K** | **L** | **M** | **N** | **O** | **P** | **Q** | **R**

S | **T** | **U** | **V** | **W** | **X** | **Y** | **Z** | **&**

2 | 3 | 4 | 5 | 6 | 7 | 8 | 9 | 0

VERTICAL LETTERING ②

①

DET.#REQ.		NAME	MAT'L	DESCRIPTION
9	1	STUD	STEEL	
8	1	DOWEL	STEEL	$\varnothing \frac{1}{8} \times \frac{1}{2}$
7	4	SOCKET HEAD CAP SCREW	STEEL	#10-24 × $\frac{1}{4}$
6	1	HEAD PRESS FIT BUSHING	STEEL	ID .3100 $^{+.0001}_{-.0005}$ × OD $\frac{9}{16} \times \frac{3}{4}$ LONG
5	1	HAND KNOB FIT BUSHING	CI	KNOB, $\varnothing \frac{5}{16}$-18 TAP
4	1	DOWEL	STEEL	
3	1	BUSHING PLATE	STEEL	
2	1	PLATE	STEEL	
1	1	BASE	STEEL	

MATERIAL	NAME
SEE DETAILS	**DRILL JIG (.316 or .154 HOLES)**

HELI-COIL
Heli-Coil Products, Div. of Mite Corp., Danbury, Conn.

BP-38

Courtesy of Heli-Coil Products, Division of Mite Corporation

SKETCHING ASSIGNMENT FOR FREEHAND VERTICAL LETTERING (BP-38)

① LETTER FREEHAND THE UPPER-CASE STRAIGHT LETTERS AND NUMERALS IN THE SPACES PROVIDED.

② LETTER FREEHAND THE VERTICAL LETTERS AND NUMERALS FOR DETAILS 1–9 OF THE PARTIAL PARTS LIST.

Freehand Inclined Lettering

Inclined letters and numbers are used on many drawings as they can be formed with a very natural movement and a slant similar to that used in everyday writing. The shape of the **inclined** or **slant** letter is the same as that of the vertical or straight letter except that circles and parts of circles are elliptical and the axis of each letter is at an angle, Figure 39–1. Slant letters may be formed with the left hand by using the same techniques of shaping and spacing as are used by the right hand. The only difference is that for some letters and numerals, the strokes are reversed.

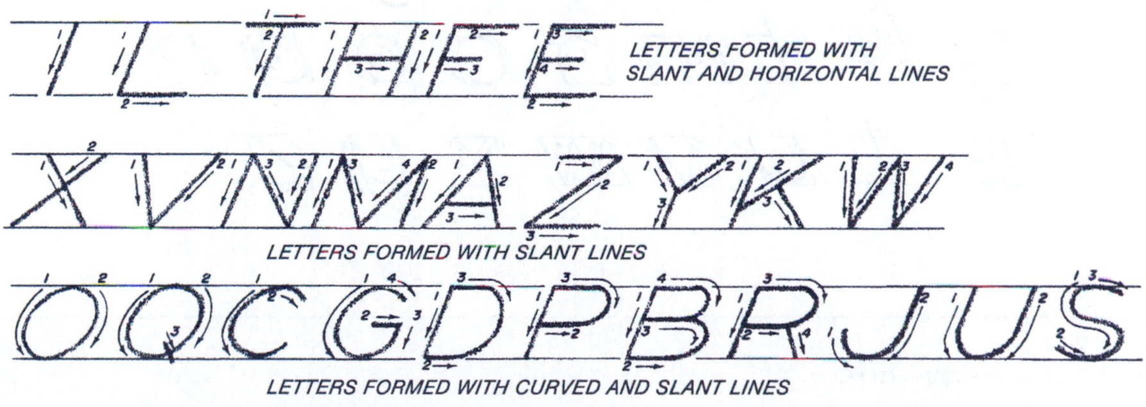

FIGURE 39–1
Forming upper-case slant letters.

FORMING SLANT LETTERS AND NUMERALS

The angle of slant of the letters and numerals may vary, depending on individual preference. Lettering at an angle from 60 to 75 degrees is practical, as it is easy to read and produce. Light horizontal guide lines to assure the uniform height of letters are recommended for the beginner. Angle guide lines assist the beginner in keeping all the letters shaped correctly and spaced properly. The direction of the strokes for each letter and numeral and the spacing between letters, words, and lines are the same as for straight letters. The shape of each upper-case slant letter is illustrated in Figure 39–1; each numeral is shown in Figure 39–2; and each lower-case slant letter is shown in Figure 39–3.

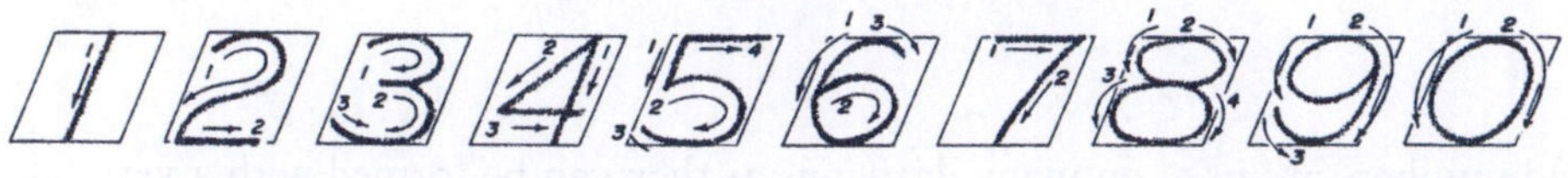

FIGURE 39–2
Forming slant numerals.

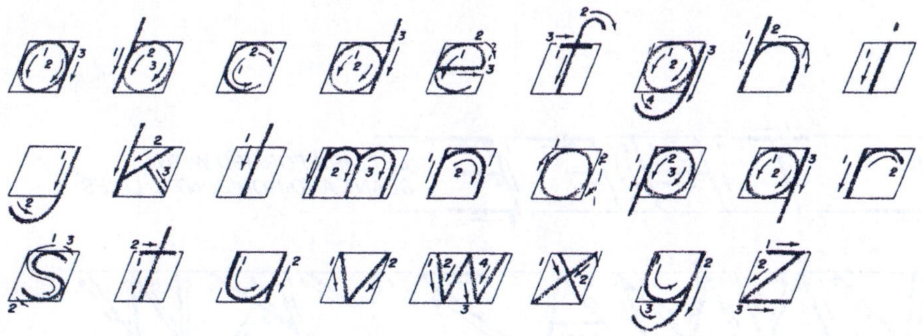

FIGURE 39–3
Forming lower-case slant letters.

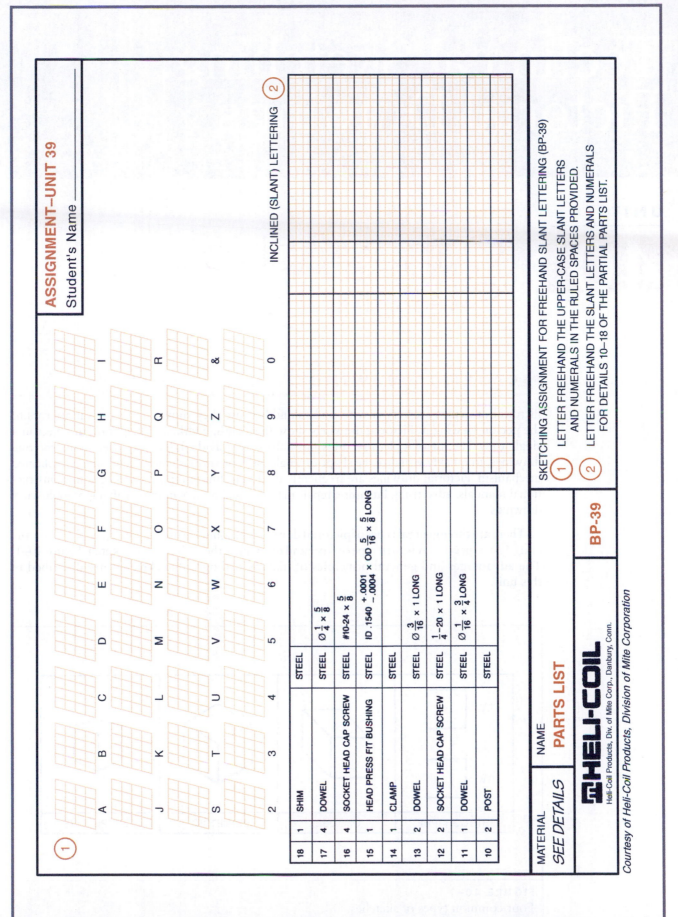

ASSIGNMENT–UNIT 39

Student's Name _____

①

	A	B	C	D	E	F	G	H	I
J	K	L	M	N	O	P	Q	R
S	T	U	V	W	X	Y	Z	&
2	3	4	5	6	7	8	9	0

② INCLINED (SLANT) LETTERING

18	1	SHIM	STEEL	
17	4	DOWEL	STEEL	$\varnothing \frac{1}{4} \times \frac{5}{8}$
16	4	SOCKET HEAD CAP SCREW	STEEL	#10-24 × $\frac{5}{8}$
15	1	HEAD PRESS FIT BUSHING	STEEL	ID .1540 $^{+.0001}_{-.0004}$ × OD $\frac{5}{16} \times \frac{5}{8}$ LONG
14	1	CLAMP	STEEL	
13	2	DOWEL	STEEL	$\varnothing \frac{3}{16} \times 1$ LONG
12	2	SOCKET HEAD CAP SCREW	STEEL	$\frac{1}{4}$–20 × 1 LONG
11	1	DOWEL	STEEL	$\varnothing \frac{1}{16} \times \frac{3}{4}$ LONG
10	2	POST	STEEL	

SKETCHING ASSIGNMENT FOR FREEHAND SLANT LETTERING (BP-39)

① LETTER FREEHAND THE UPPER-CASE SLANT LETTERS AND NUMERALS IN THE RULED SPACES PROVIDED.

② LETTER FREEHAND THE SLANT LETTERS AND NUMERALS FOR DETAILS 10–18 OF THE PARTIAL PARTS LIST.

| MATERIAL SEE DETAILS | NAME PARTS LIST |

⧟ HELI-COIL

Heli-Coil Products, Div. of Mite Corp., Danbury, Conn.

BP-39

Courtesy of Heli-Coil Products, Division of Mite Corporation

UNIT 39 | FREEHAND INCLINED LETTERING

Technical Sketching: Pictorial Drawings

40 Orthographic Sketching

PICTORIAL DRAWINGS

Pictorial drawings are very easy to understand because they show an object as it appears to the person viewing it. Pictorial drawings show the length, width, and height of an object in a single view. The use of a pictorial drawing enables an individual, inexperienced in interpreting drawings, to quickly visualize the shape of single parts or various components in a complicated mechanism. Pictorial drawings are frequently used in technical reports, parts catalogs, instructional manuals, advertising literature, repair manuals, and as aids in the clarifying of multi-view drawings.

There are three general types of pictorial drawings in common use: ① oblique, ② isometric, and ③ perspective. A fourth type of freehand drawing is the orthographic sketch, Figure 40–1. The advantages and general principles of making orthographic sketches are described in this unit.

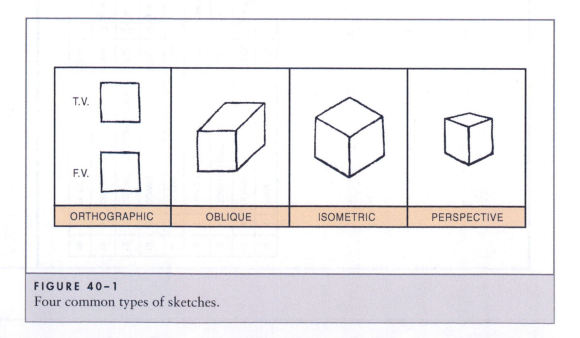

| ORTHOGRAPHIC | OBLIQUE | ISOMETRIC | PERSPECTIVE |

T.V.

F.V.

FIGURE 40–1
Four common types of sketches.

MAKING ORTHOGRAPHIC SKETCHES

The orthographic sketch is the simplest type to make of the four types of sketches. The views (Figure 40–2) are developed as in any regular instrument drawing, and the same types of lines are used. The only difference is that orthographic sketches are drawn freehand. In some situations, orthographic sketches are made with sufficient quality and quantity of information that they replace the need for a mechanical or computer drawing.

This is rarely possible with pictorial sketches covered in the next three units. In actual practice, on-the-spot sketches of small parts are made as near the actual size as possible.

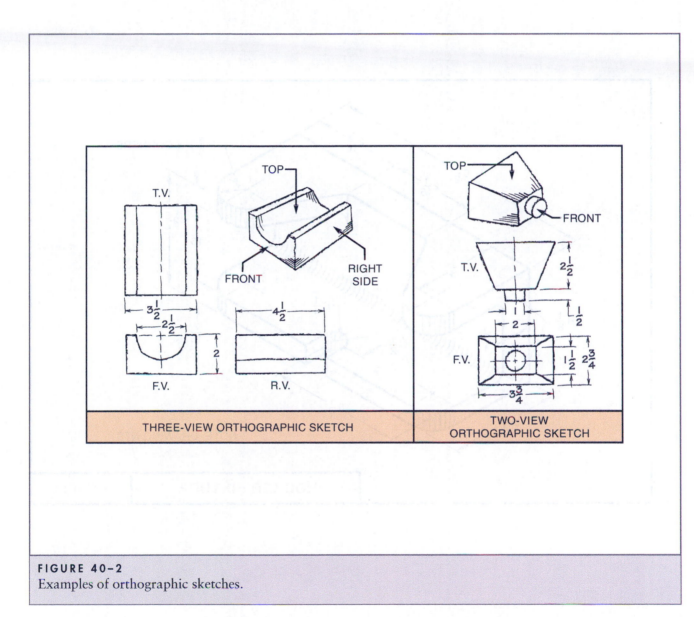

THREE-VIEW ORTHOGRAPHIC SKETCH | TWO-VIEW ORTHOGRAPHIC SKETCH

FIGURE 40–2
Examples of orthographic sketches.

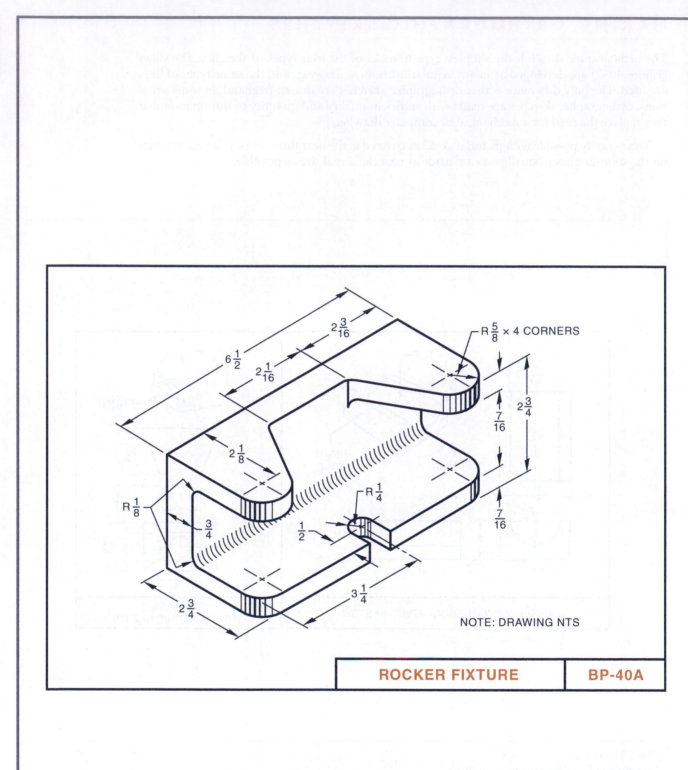

$R\frac{5}{8} \times 4$ CORNERS

$2\frac{3}{16}$

$6\frac{1}{2}$

$2\frac{1}{16}$

$2\frac{3}{4}$

$2\frac{1}{8}$

$\frac{7}{16}$

$R\frac{1}{4}$

$R\frac{1}{8}$

$\frac{7}{16}$

$\frac{3}{4}$

$\frac{1}{2}$

$2\frac{3}{4}$

$3\frac{1}{4}$

NOTE: DRAWING NTS

ROCKER FIXTURE   **BP-40A**

ASSIGNMENT A–UNIT 40

Student's Name _____

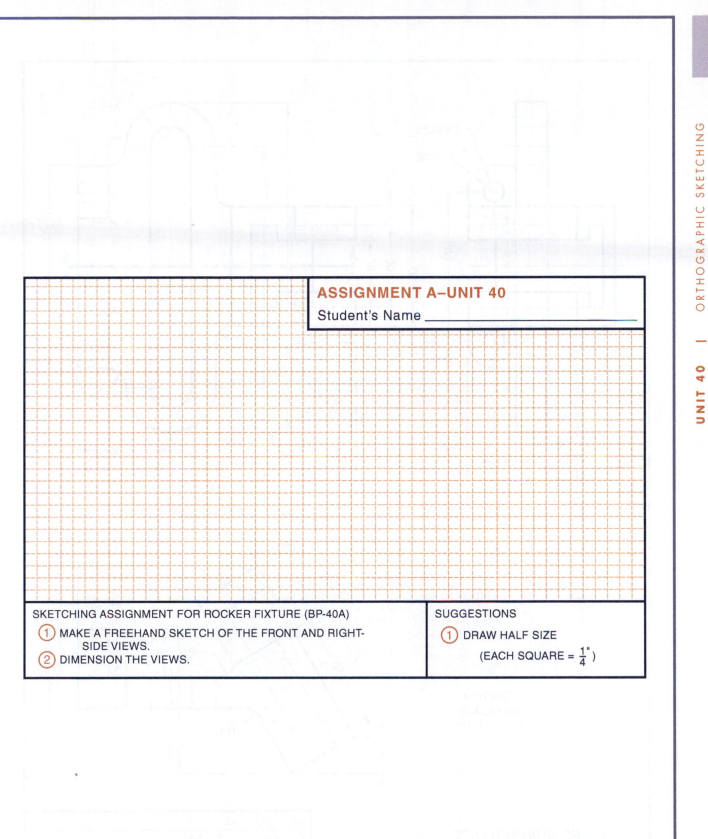

SKETCHING ASSIGNMENT FOR ROCKER FIXTURE (BP-40A)

1 MAKE A FREEHAND SKETCH OF THE FRONT AND RIGHT-
   SIDE VIEWS.
2 DIMENSION THE VIEWS.

SUGGESTIONS

1 DRAW HALF SIZE
   (EACH SQUARE = $\frac{1}{4}$")

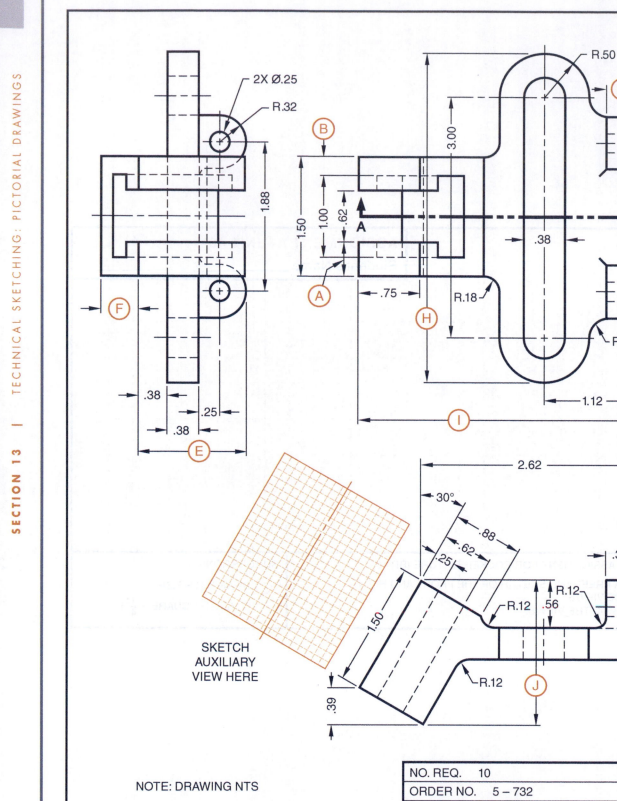

2X Ø.25

R.32

1.88

B

A

.38

.25

.38

E

F

B

1.50

1.00

.62

A

.75

H

3.00

R.50

G

C

.38

R.18

R.18

1.12

I

A

A

SKETCH
AUXILIARY
VIEW HERE

2.62

30°

.88

.62

.25

1.50

.39

R.12

R.12

R.12

.56

.38

D

J

NOTE: DRAWING NTS

NO. REQ.	10
ORDER NO.	5 – 732
MATL.	CAST STEEL
GUIDE BRACKET	**BP-40B**

ASSIGNMENT B1—UNIT 40: GUIDE BRACKET (BP-40B)

Student's Name _____

Complete assignment B1, Unit 40. Use answers to complete assignment B2, Unit 40.

1. _____

_____

_____

1. Name each of the three views.

2. Give the number and diameter of the drilled holes.

2. No. _____ Dia _____

3. Determine dimensions (A) and (B).

3. (A) = _____ (B) = _____

4. Compute dimension (C).

4. (C) = _____

5. What is the dimension of the leg (D)?

5. (D) = _____

6. Determine dimension (E).

6. (E) = _____

7. Give dimension (F) and (G).

7. (F) = _____ (G) = _____

8. What is the overall height of (H)?

8. (H) = _____

9. Determine overall length of (I).

9. (I) = _____

10. Give overall depth of (J).

10. (J) = _____

ASSIGNMENT B2—UNIT 40

Student's Name _____

SKETCHING ASSIGNMENT FOR GUIDE BRACKET (BP-40B)   NOTE: 1 SQ. = .10" □

(1) SKETCH FREEHAND THE REQUIRED AUXILIARY VIEW IN THE SPACE PROVIDED ON BP-40B.

(2) SKETCH FREEHAND AND DIMENSION FULL SECTION VIEW A-A ON ABOVE GRID.

Oblique Sketching

An oblique sketch is a type of pictorial drawing on which two or more surfaces are shown at one time on one drawing. The front face of the object is sketched in the same manner as the front view of either an orthographic sketch or a mechanical drawing. All of the straight, inclined, and curved lines on the front plane of the object will appear in their true size and shape on this front face. Because the other sides of the object are sketched at an angle, the surfaces and lines are not shown in their true size and shape. Oblique sketches are frequently used in cabinet making, woodworking, and construction.

MAKING OBLIQUE SKETCHES WITH STRAIGHT LINES

The steps in making an oblique sketch are simple. For example, if an oblique sketch of a rectangular die block is needed, seven basic steps are followed, as shown in Figure 41–1.

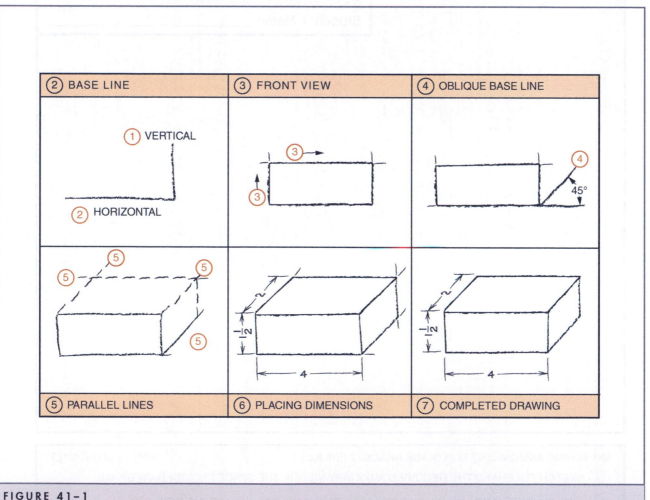

FIGURE 41-1
Steps in making an oblique sketch with dimensions.

STEP 1 Select one view of the object that gives most of the desired information.

STEP 2 Draw light horizontal and vertical baselines.

STEP 3 Lay out the edges of the die block in the front view from the base lines. All vertical lines on the front face of the object will be parallel to the vertical base line; all horizontal lines will be parallel to the horizontal base line. All lines will be in their true size and shape in this view.

STEP 4 Start at the intersection of the vertical and horizontal base lines and draw a line at an angle of 45° to the base line. This line is called the oblique base line.

STEP 5 Draw the remaining lines for the right-side and top view parallel to either the oblique base line or to the horizontal or vertical base lines, as the case may be. Because these lines are not in their true size or shape, they should be drawn so they appear in proportion to the front view.

STEP 6 Place dimensions so they are parallel to the edge lines.

STEP 7 Erase unnecessary lines. Darken object lines to make the sketch clearer and easier to interpret.

SKETCHING CIRCLES IN OBLIQUE

A circle or arc located on the front face of an oblique sketch is drawn in its true size and shape. However, because the top and side views are distorted, a circle or arc will be elliptical in these views. Three circles drawn in oblique on the top, right-side, and front of a cube are shown in Figure 41–2.

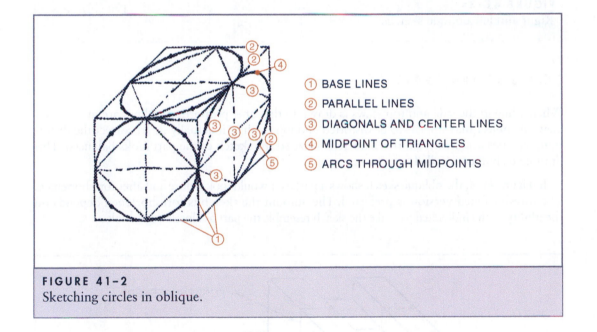

① BASE LINES

② PARALLEL LINES

③ DIAGONALS AND CENTER LINES

④ MIDPOINT OF TRIANGLES

⑤ ARCS THROUGH MIDPOINTS

FIGURE 41–2
Sketching circles in oblique.

STEP 1 Draw horizontal, vertical, and oblique base lines.

STEP 2 Sketch lines parallel to these base lines or axes to form a cube.

STEP 3 Draw center lines and diagonals in the front, right-side, and top faces.

STEP 4 Locate the midpoints of triangles formed in the three faces.

STEP 5 Draw curved lines through these points. NOTE: The circle appears in its true size and shape in the front face and as an ellipse in the right-side and top faces.

STEP 6 Touch up and darken the curved lines. Erase guidelines where they are of no value in reading the sketch.

RIGHT AND LEFT OBLIQUE SKETCHES

Up to this point, the object has been viewed from the right-side. In many cases, the left-side of the object must be sketched because it contains better details. In these instances, the oblique base line or axis is 45° from the horizontal base line, starting from the left-edge of the object. A rectangular die block sketched from the right-side is shown in Figure 41–3A. The same rectangular die block is shown in Figure 41–3B as it would appear when sketched from the left-side.

The same principles and techniques for making oblique sketches apply, regardless of whether the object is drawn in the right or left position.

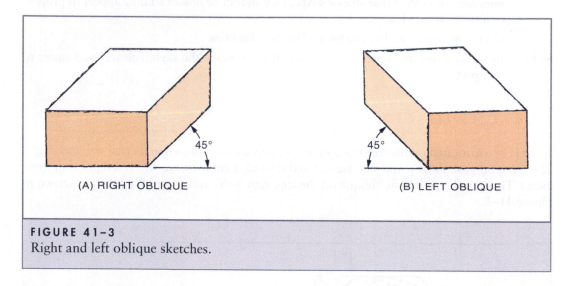

(A) RIGHT OBLIQUE 45°   45° **(B) LEFT OBLIQUE**

FIGURE 41–3
Right and left oblique sketches.

FORESHORTENING

When a line in the side and top views is drawn in the same proportion as lines are in the front view, the object may appear to be distorted and longer than it actually is. To correct the distortion, the lines are drawn shorter than actual size so the sketch of the part looks balanced. This drafting technique is called foreshortening.

In Figure 41–4, the oblique sketch shows a part as it would look before and after foreshortening. The foreshortened version is preferred. The amount the sketch is foreshortened depends on the ability of the individual to make the sketch resemble the part as closely as possible.

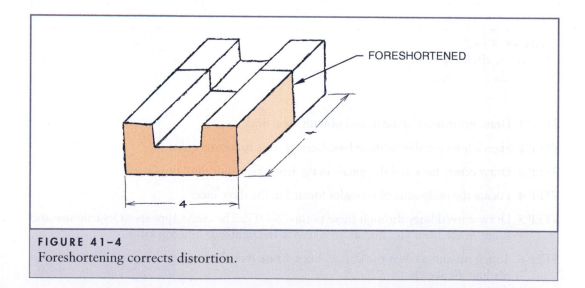

FORESHORTENED

4

FIGURE 41–4
Foreshortening corrects distortion.

ASSIGNMENT–UNIT 41

Student's Name _____

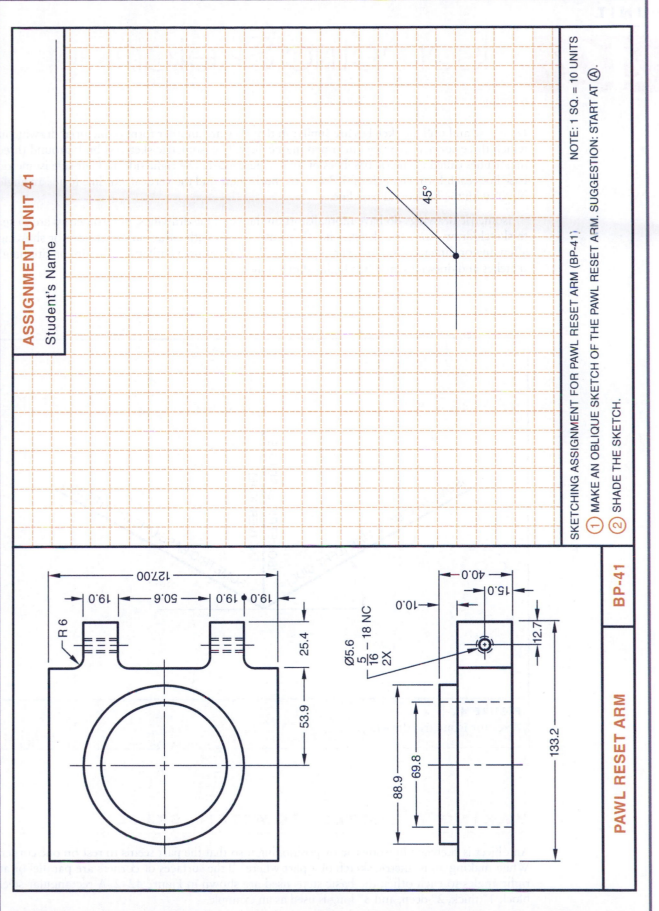

45°

NOTE: 1 SQ. = 10 UNITS

SKETCHING ASSIGNMENT FOR PAWL RESET ARM (BP-41)

① MAKE AN OBLIQUE SKETCH OF THE PAWL RESET ARM. SUGGESTION: START AT Ⓐ.

② SHADE THE SKETCH.

PAWL RESET ARM   **BP-41**

127.00

19.0   50.6   19.0   19.0

R 6

25.4

53.9

Ø5.6
5/16 – 18 NC
2X

40.0

15.0

10.0

12.7

88.9

69.8

133.2

42 Isometric Sketching

Isometric and oblique sketches are similar in that they are another form of pictorial drawing in which three surfaces may be illustrated in one view. The isometric sketch is built around three major lines called isometric baselines or axes, Figure 42–1. The right-side and left-side isometric base lines each form an angle of 30° with the horizontal, and the vertical axis line forms an angle of 90° with the horizontal base line.

Isometric drawings do have more distortion than oblique drawings, but isometrics are easier to sketch/draw. Isometric sketches are frequently drawn with either no dimensions or showing a few major dimensions. However, piping & HVAC (heating, ventilating and air-conditioning) sketches and drawings are often highly dimensioned.

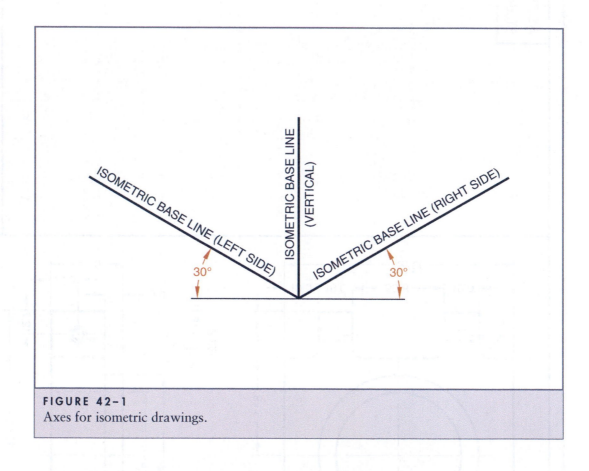

FIGURE 42–1
Axes for isometric drawings.

MAKING A SIMPLE ISOMETRIC SKETCH

An object is sketched in isometric by positioning it so that the part seems to rest on one corner. When making an isometric sketch of a part where all the surfaces or corners are parallel or at right angles to each other, the basic steps used are shown in Figure 42–2. A rectangular steel block 1" thick, 2" deep, and 3" long is used as an example.

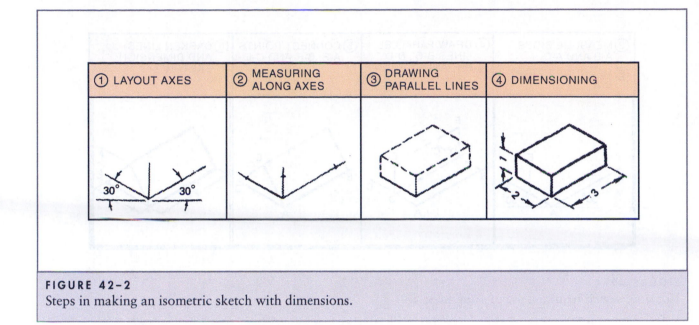

① LAYOUT AXES	② MEASURING ALONG AXES	③ DRAWING PARALLEL LINES	④ DIMENSIONING

FIGURE 42–2
Steps in making an isometric sketch with dimensions.

STEP 1 Sketch the three isometric axes. If a ruled isometric sheet is available, select three lines for the major axes.

STEP 2 Lay off the 3″ length along the right axis, the 2″ depth on the left axis line, and the 1″ height on the vertical axis line.

STEP 3 Draw lines from these layout points parallel to the three axes. Note that all parallel lines on the object are parallel on the sketch.

STEP 4 Dimension the sketch. On isometric sketches, the dimensions are placed parallel to the edges.

SKETCHING SLANT LINES IN ISOMETRIC

Only those lines that are parallel to the axes may be measured in their true lengths. Slant lines representing inclined surfaces are not shown in their true lengths in isometric sketches. In most cases, the slant lines for an object (such as the casting shown in Figure 42–3) may be drawn by following the steps shown in Figure 42–4.

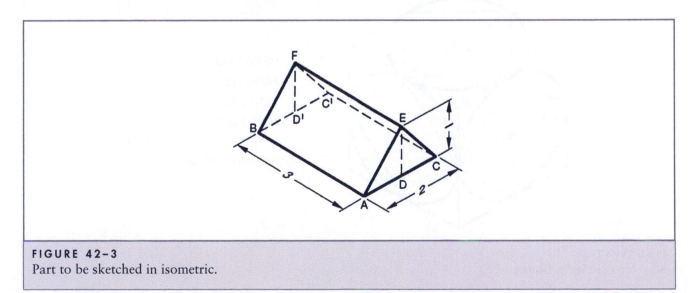

FIGURE 42–3
Part to be sketched in isometric.

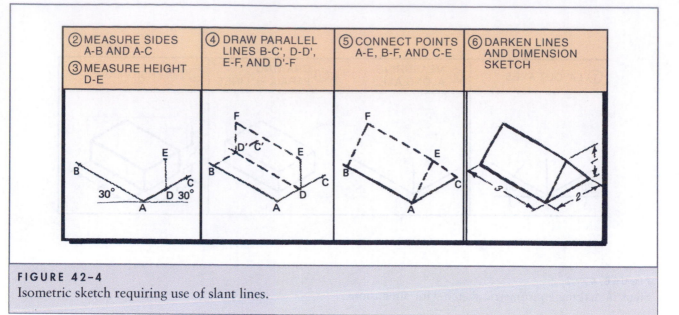

②MEASURE SIDES A-B AND A-C ③MEASURE HEIGHT D-E	④DRAW PARALLEL LINES B-C', D-D', E-F, AND D'-F	⑤CONNECT POINTS A-E, B-F, AND C-E	⑥DARKEN LINES AND DIMENSION SKETCH

FIGURE 42–4
Isometric sketch requiring use of slant lines.

STEP 1 Draw the three major isometric axes.

STEP 2 Measure distance A-B on the left axis and A-C on the right axis.

STEP 3 Measure distance A-D on the right axis and draw a vertical line from this point. Lay out distance D-E on this line.

STEP 4 Draw parallel lines B-C', D-D', E-F, and D'-F.

STEP 5 Connect points A-E, B-F, and C-E.

STEP 6 Darken lines and dimension the isometric sketch.

SKETCHING CIRCLES AND ARCS IN ISOMETRIC

The techniques used to sketch arcs and circles in oblique also may be used for sketching arcs and circles in isometric. Each step is illustrated in Figure 42–5. Note that in each face, the circle appears as an ellipse. This is also shown in Figure 42–6.

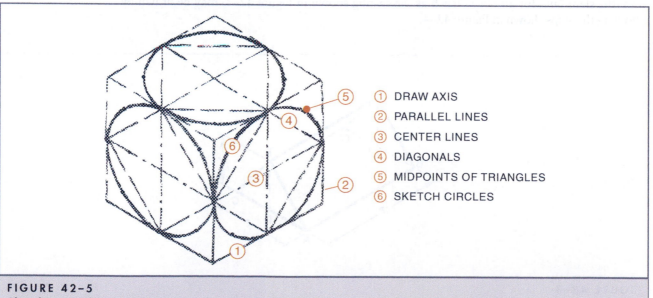

① DRAW AXIS
② PARALLEL LINES
③ CENTER LINES
④ DIAGONALS
⑤ MIDPOINTS OF TRIANGLES
⑥ SKETCH CIRCLES

FIGURE 42–5
Sketching circles in isometric.

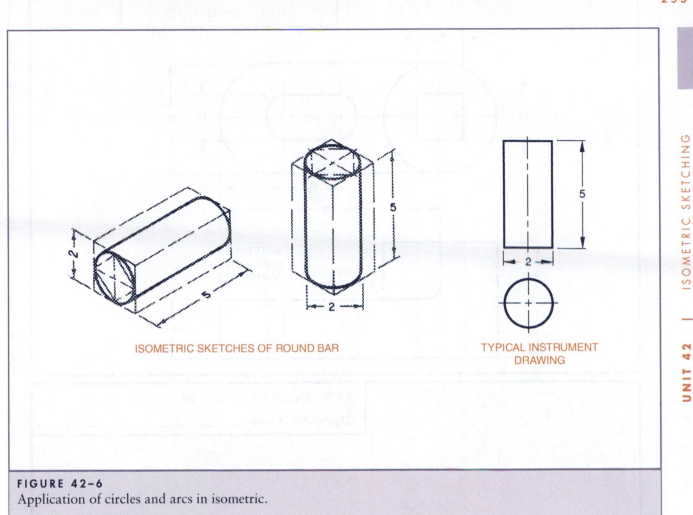

ISOMETRIC SKETCHES OF ROUND BAR

TYPICAL INSTRUMENT
DRAWING

FIGURE 42-6
Application of circles and arcs in isometric.

LAYOUT SHEETS FOR ISOMETRIC SKETCHES

The making of isometric drawings can be simplified if specially ruled isometric layout sheets
are used. Considerable time may be saved as these graph sheets provide guide lines that run
in three directions. The guide lines are parallel to the three isometric axes and to each other.
The diagonal lines of the graph paper usually are at a given distance apart to simplify the
making of the sketch and to ensure that all lines are in proportion to the actual size. The
ruled lines are printed on a heavy paper that can be used over and over again as an underlay
sheet for tracing paper.

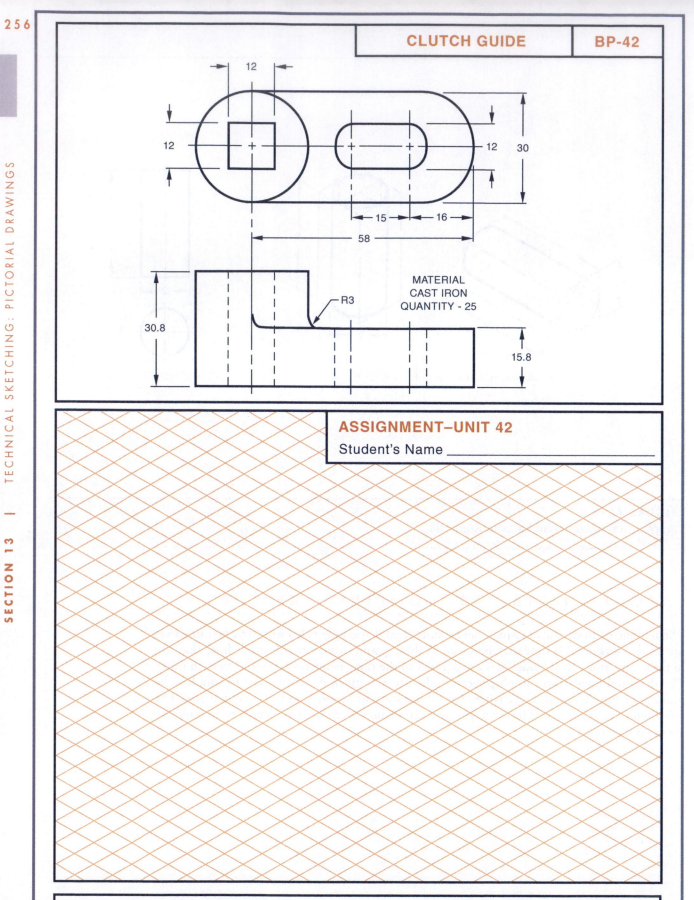

CLUTCH GUIDE | **BP-42**

MATERIAL
CAST IRON
QUANTITY - 25

ASSIGNMENT—UNIT 42

Student's Name _____

SKETCHING ASSIGNMENT FOR CLUTCH GUIDE (BP-42)   NOTE: 1 GRID SQ. = 5MM □

1 MAKE A FREEHAND ISOMETRIC SKETCH OF THE CLUTCH GUIDE.

2 SHADE.

3 DIMENSION THE SKETCH COMPLETELY.

Perspective Sketching

Perspective sketches show two or three sides of an object in one view. As a result, they resemble a photographic picture. In both the perspective sketch and the photograph, the portion of the object that is closest to the observer is the largest. The parts that are farthest away are smaller. Lines and surfaces on perspective sketches become smaller and come closer together as the distance from the eye increases. Eventually, they seem to disappear at an imaginary horizon.

A perspective drawing is best used when large objects are to be seen from a distance with minimal distortion. Like other pictorial sketches, detailed dimensioning is difficult.

SINGLE-POINT PERSPECTIVE

There are two types of perspective sketches commonly used. The first type is known as single-point or parallel perspective. In this case, one face of the object in parallel perspective is sketched in its true size and shape, the same as in an orthographic sketch. For example, the edges of the front face of a cube, as shown in parallel perspective in Figure 43–1 at Ⓐ, Ⓑ, Ⓒ, and Ⓓ, are parallel and square.

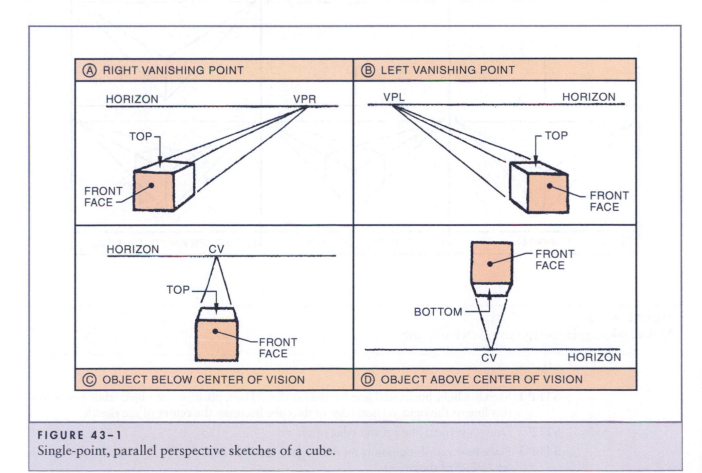

FIGURE 43–1
Single-point, parallel perspective sketches of a cube.

To draw the remainder of the cube, the two sides decrease in size as they approach the horizon. On this imaginary horizon, which is supposed to be at eye level, the point where the lines come together is called the **vanishing point.** This vanishing point may be above or below the object or to the right or left of it, Figure 43–1. Vertical lines on the object are vertical on parallel perspective sketches and do not converge. Single point parallel perspective is the simplest type of perspective to understand and the easiest to sketch.

Parallel prospective CAD drawings are used to show 3-D objects in a dimensionally correct manner. Relative and scaled dimensions may be taken from the drawings. The down side is that parallel perspective drawings do not appear as realistic as isometric or angular perspective drawings.

ANGULAR OR TWO-POINT PERSPECTIVE

The second type of perspective sketch is the two-point perspective, also known as **angular perspective.** As the name implies, two vanishing points on the horizon are used and all lines converge toward these points. Usually, in a freehand perspective sketch, the horizon is in a horizontal position. Seven basic steps are required to make a two-point perspective of a cube, using right and left vanishing points. The application of each of these steps is shown in Figure 43–2.

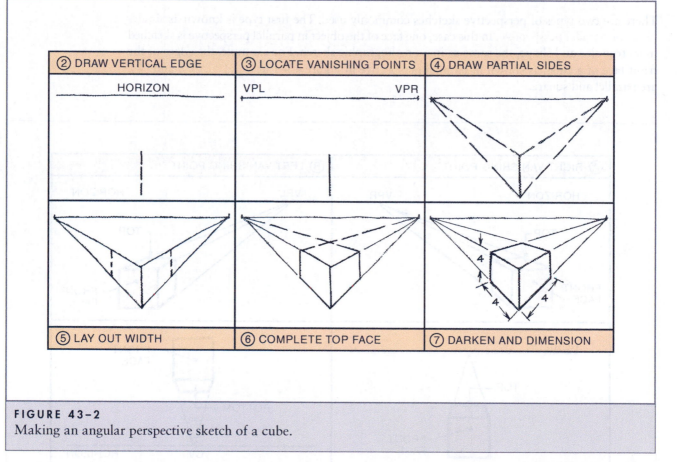

FIGURE 43–2
Making an angular perspective sketch of a cube.

STEP 1 Sketch a light horizontal line for the **horizon.** Then, position the object above or below this line so the right vertical edge of the cube becomes the center of the sketch.

STEP 2 Draw a vertical line for this edge of the cube.

STEP 3 Place two vanishing points on the horizon: one to the right **(VPR)** and the other to the left **(VPL)** of the object.

STEP 4 Draw light lines from the corners of the vertical line to the vanishing points.

STEP 5 Lay out the width of the cube and then draw parallel vertical lines.

STEP 6 Sketch the two remaining lines for the top, starting at the points where the two vertical lines intersect the top edge of the cube.

STEP 7 Darken all object lines and dimension. Once again, each dimension is placed parallel to the edge that it measures.

SKETCHING CIRCLES AND ARCS IN PERSPECTIVE

Circles and arcs are distorted in all views of perspective drawings except in one face of a parallel perspective drawing. Circles are drawn in perspective by using the same techniques of blocking-in that apply to orthographic, oblique, and isometric sketches. The steps in drawing circles in perspective are summarized in Figure 43–3. The same practices may be applied to sketching arcs.

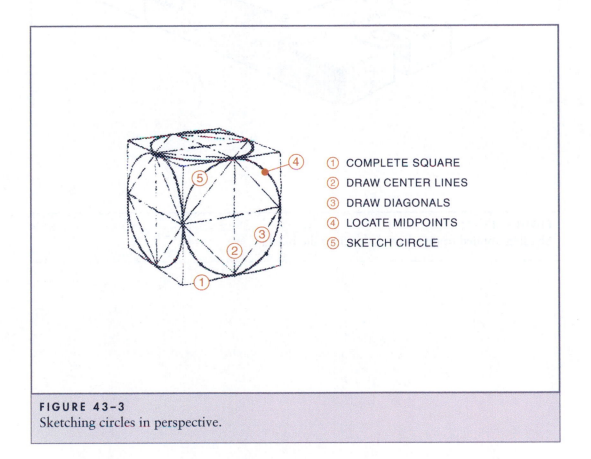

① COMPLETE SQUARE
② DRAW CENTER LINES
③ DRAW DIAGONALS
④ LOCATE MIDPOINTS
⑤ SKETCH CIRCLE

FIGURE 43–3
Sketching circles in perspective.

STEP 1 Lay out the four sides of the square that correspond to the diameter of the required circle. Note that two of the sides converge toward the vanishing point.

STEP 2 Draw the center lines.

STEP 3 Draw the diagonals.

STEP 4 Locate the midpoints of the triangles that were formed.

STEP 5 Sketch the circle through the points where the center lines touch the sides of the square and the midpoints.

SHADING PERSPECTIVE SKETCHES

Many perspective sketches need to create the illusion of depth. This can be accomplished by using tonal line shading. Applying overlapping penciled lines creates darker tonal areas to lighter tonal areas, Figure 43–4. Also, the same techniques used in shading oblique and isometric sketches also apply to perspectives.

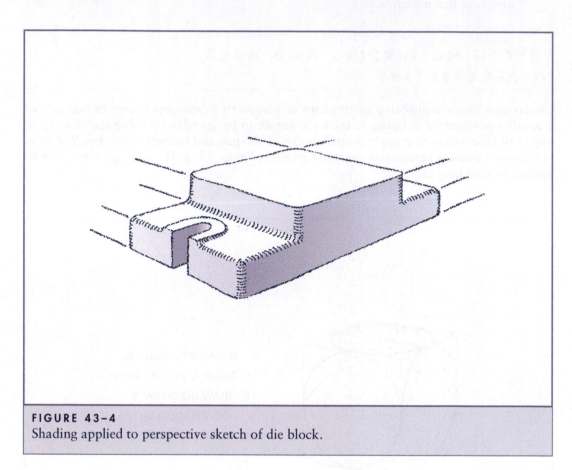

FIGURE 43–4
Shading applied to perspective sketch of die block.

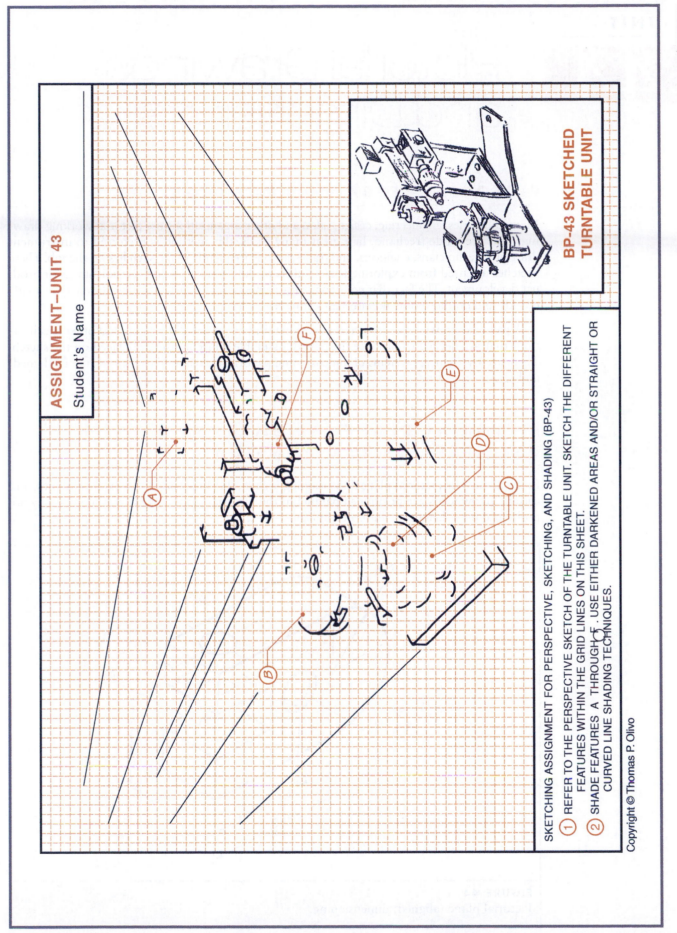

ASSIGNMENT–UNIT 43

Student's Name

BP-43 SKETCHED TURNTABLE UNIT

SKETCHING ASSIGNMENT FOR PERSPECTIVE, SKETCHING, AND SHADING (BP-43)

① REFER TO THE PERSPECTIVE SKETCH OF THE TURNTABLE UNIT. SKETCH THE DIFFERENT FEATURES WITHIN THE GRID LINES ON THIS SHEET.

② SHADE FEATURES A THROUGH F. USE EITHER DARKENED AREAS AND/OR STRAIGHT OR CURVED LINE SHADING TECHNIQUES.

Pictorial Drawings and Dimensions

PICTORIAL WORKING DRAWINGS

Many industries, engineering, design, sales, and other organizations use pictorial working drawings that are made freehand. In a great many cases, this method is preferred to instrument drawings. Technicians, engineers, and tradespersons frequently make sketches on the job. These sketches are used from exploring project ideas to the completion and revision stages of product development. The fact that no drawing instruments are required and sketches may be made quickly and on-the-spot makes freehand drawings practical.

A pictorial drawing is considered a working drawing when dimensions and other specifications that are needed to produce the part or assemble a mechanism are placed on the sketch. There are two general systems of dimensioning pictorial sketches: (1) **pictorial plane** (aligned) and (2) **unidirectional.**

PICTORIAL PLANE (ALIGNED) AND UNIDIRECTIONAL DIMENSIONS

In the aligned or pictorial plane dimensioning system, the dimension lines, extension lines, and certain arrowheads are positioned parallel to the pictorial planes. These features, as they are positioned on a sketch, are illustrated in Figure 44–1.

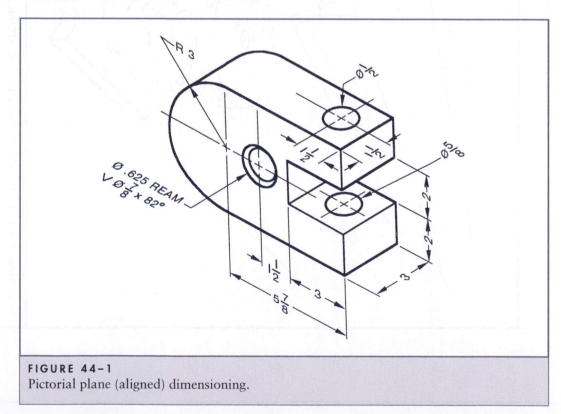

FIGURE 44–1
Pictorial plane (aligned) dimensioning.

By contrast, dimensions, notes, and technical details are lettered vertically in unidirectional dimensioning, Figure 44–2. In this vertical position, the dimensions are easier to read.

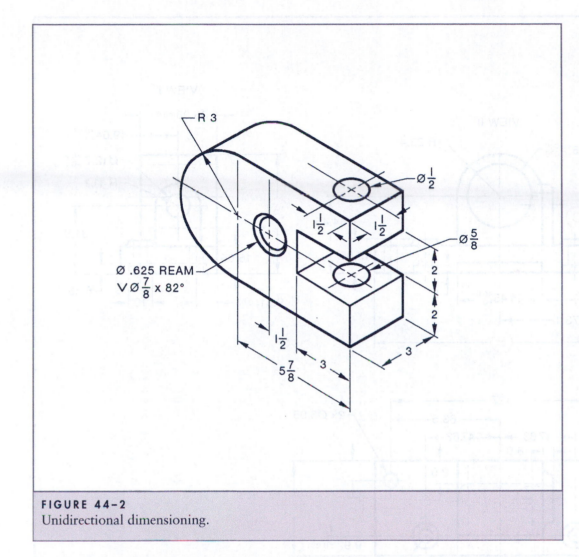

FIGURE 44–2
Unidirectional dimensioning.

Pictorials can be prepared from multiview drawings or vice versa (multiviews can be prepared from pictorials). Thus, the dimensions, notes, and important data included on the sketch should conform to the "standards" of the projection system being used.

PICTORIAL DIMENSIONING RULES

Essentially, the same basic rules for dimensioning a pictorial sketch are followed as for a multiview drawing. The basic pictorial dimensioning rules apply to both the aligned and unidirectional dimensioning systems.

STEP 1 Dimension and extension lines are drawn parallel to the pictorial planes.

STEP 2 Dimensions are placed on visible features whenever possible.

STEP 3 Arrowheads lie in the same plane as extension and dimension lines.

STEP 4 Notes and dimensions are lettered parallel with the horizontal plane.

Because a pictorial drawing is a one-view drawing, it is not always possible to avoid dimensioning on the object, across dimension lines, or on hidden surfaces. Effort should be made to avoid these practices. This will prevent errors in reading dimensions or interpreting particular features of the part(s).

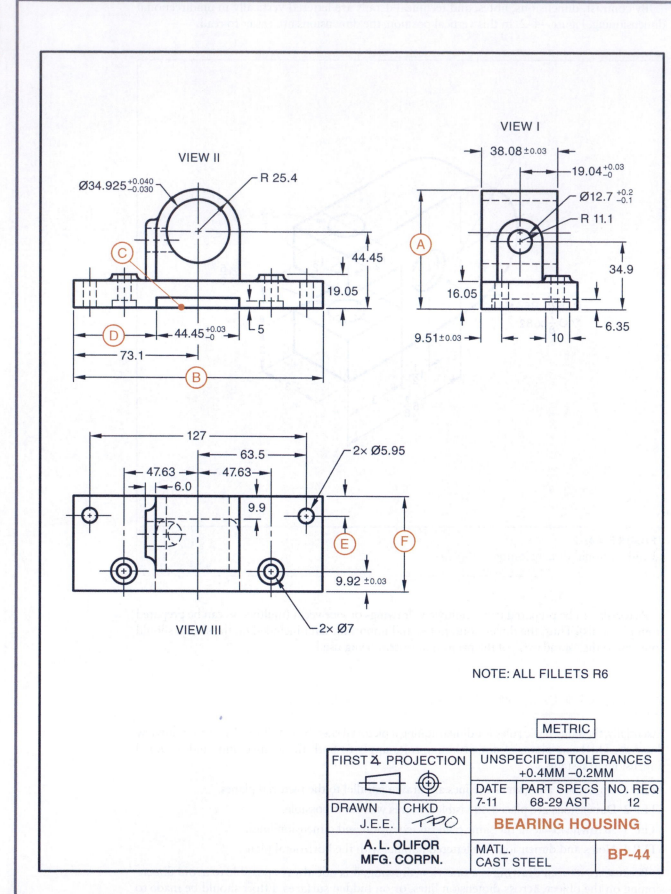

VIEW II

Ø34.925 $^{+0.040}_{-0.030}$

R 25.4

C

44.45

19.05

D

44.45 $^{+0.03}_{-0}$

5

73.1

B

VIEW I

38.08 ±0.03

19.04 $^{+0.03}_{-0}$

Ø12.7 $^{+0.2}_{-0.1}$

R 11.1

A

34.9

16.05

9.51 ±0.03

10

6.35

127

63.5

47.63

47.63

6.0

2× Ø5.95

9.9

E

F

9.92 ±0.03

2× Ø7

VIEW III

NOTE: ALL FILLETS R6

METRIC

FIRST ∠ PROJECTION	UNSPECIFIED TOLERANCES +0.4MM −0.2MM		
	DATE 7-11	PART SPECS 68-29 AST	NO. REQ 12
DRAWN J.E.E.	CHKD TPO	**BEARING HOUSING**	
A. L. OLIFOR MFG. CORPN.	MATL. CAST STEEL		**BP-44**

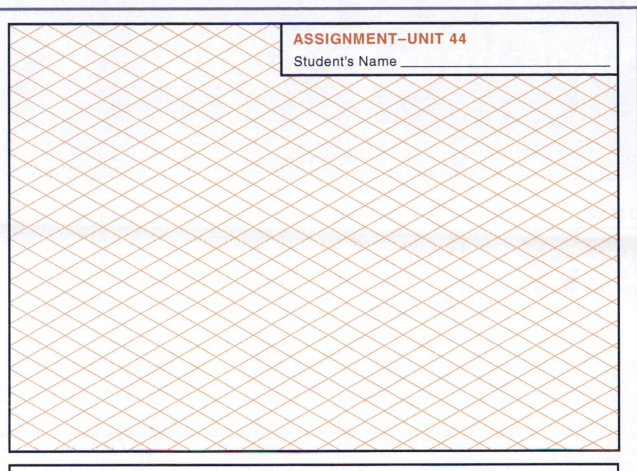

ASSIGNMENT–UNIT 44

Student's Name _____

SKETCHING ASSIGNMENT FOR BEARING HOUSING (BP-44)   NOTE: 1 SQ. = 10MM □

① MAKE A FREEHAND ISOMETRIC SKETCH OF THE BEARING HOUSING.

② DIMENSION THE SKETCH COMPLETELY. USE METRIC DIMENSIONS AND THE PICTORIAL PLANE (ALIGNED) DIMENSIONING SYSTEM.

ASSIGNMENT—UNIT 44: BEARING HOUSING (BP-44)

Student's Name _____

1. State what angle of projection is used.

2. Name VIEWS I, II, and III.

3. Compute maximum overall height Ⓐ.

4. Determine the lower-limit of Ⓑ.

5. Determine the maximum height and depth of slot Ⓒ.

6. Determine the maximum distance Ⓓ.

7. Compute the minimum and maximum center line distance Ⓔ.

8. Compute the maximum overall depth Ⓕ.

9. What does the oval hole in VIEW III represent?

10. Give the specifications for the counterbored holes.

1. _____

2. I = _____ II = _____

   III = _____

3. Ⓐ = _____

4. Ⓑ = _____

5. Ⓒ = _____

   _____

6. Ⓓ = _____

7. Ⓔ = _____

8. Ⓕ = _____

9. _____

10. _____

Sketching for CAD/CNC

45

Two-Dimensional and Three-Dimensional CAD Sketching

Both CNC machines and CADD (Computer-Aided Drafting and Design) use two- and three-dimensional Cartesian coordinates in the design and manufacturing of detailed and assembled structures.

In addition to understanding the Cartesian X, Y, and Z axes (Unit 28) and Datums (Unit 26), the CAD/CADD/CNC drawing must take into account the zero origin point. This point is called the machine zero. From this point, all machine movement starts to take place. Many CNC machines use a "G" code programming to define a tool path and speed to draw or make a part.

In 2007, a new machine tool control language was developed to enhance the older G-code. It is a language that connects CNC process data with the product description of the product being machined. Called STEP-NC, it allows tool parts descriptions to be independent of the machine tool. It also allows parts inspection information to be passed back directly to the machine tool to correct process errors.

Two methods of indicating movement to points or positions are used: absolute and incremental. Absolute coordinates use the same concept for defining 2-D and 3-D points.

The absolute method defines a point from machine zero. Example: Point A (Figure 45–1) would be located at 2,2,3; point B would be located at 5,2,3. The incremental movement defines a point from the machine's current position. In this example, to move from point A to point B, the command would be 3,0,0.

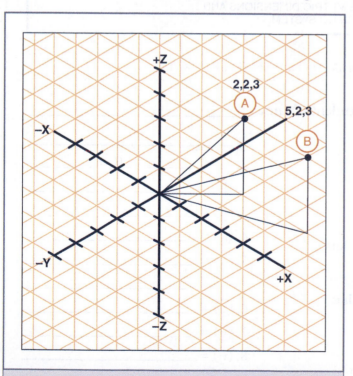

FIGURE 45–1
Incremental and absolute coordinates.

Both wire-frame and solid 3-D objects can be sketched by using these dimensioning concepts.

A wire-frame model appears to be transparent, using lines to represent all edges. Hidden edges are indicated with broken lines. (See Figure 45–2 for a partial wire-frame model.)

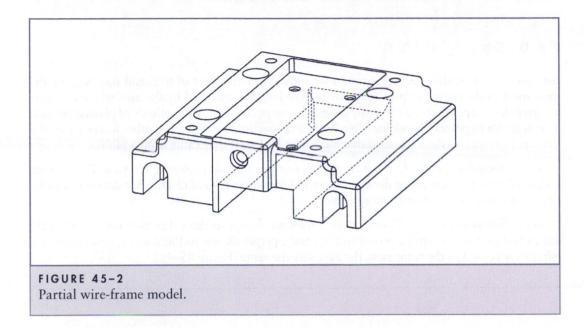

FIGURE 45–2
Partial wire-frame model.

A surface model can be described as a wire-frame model with a sheet of material stretched over it (Figure 45–3). In a solid model, no hidden surfaces or edges are shown.

Sketches of 2-D and 3-D models are frequently used in the "idea" stages of product development and revision. By sketching a mental idea on paper, it can be communicated to others and adjustments incorporated.

In addition, equipment called Stereolithography or 3-D printing can speed up the manufacturing process by creating a physical prototype directly from a CAD 3-D model. The Stereolithography printer creates a 3-D model by slicing the 3-D CAD data into thin cross sections and modulates a laser that creates a 0.005″ thick layer of photosensitive liquid epoxy plastic. The printer builds the prototype layer by layer from the bottom up.

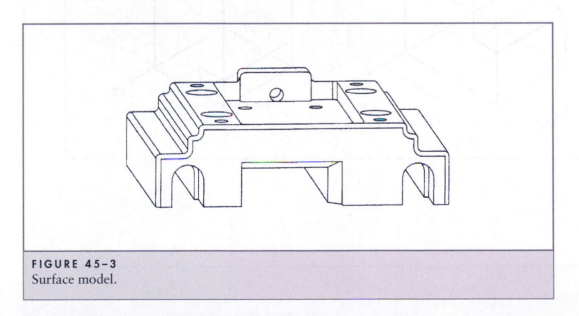

FIGURE 45–3
Surface model.

A second type of rapid prototyping called solid object 3-D printing uses a similar technique that ink jet printers use. The print head with many tiny jets builds models by dispensing a thermoplastic material in layers.

Prototype models made with these techniques are used for design verification, sales presentations, investment casting, tooling, and other manufacturing functions.

CAD SKETCHING

An option exists within CAD programs that allows the equivalent of freehand sketching by the movement of the mouse or puck. The created line length is specified by the user, who can generate irregular shapes that are not easily created by more conventional methods of placing entities. This is useful to portray break lines in a broken-out sectional view or any other features that the draftsperson may wish to create so that they don't look as if they came from a plotter.

Two-dimensional (2-D) CAD drawings appear as a normal orthographic view. Every point is defined by a position along the x and y axes. Three-dimensional (3-D) CAD drawings can be drawn in either of two coordinate systems.

In the first system, View Coordinates, values are drawn so the x direction is the horizontal axis, y is drawn in the vertical axis direction, and z perpendicular to the x and y axis. No matter what view is used as the front view, the axes stay the same (Figure 45–4).

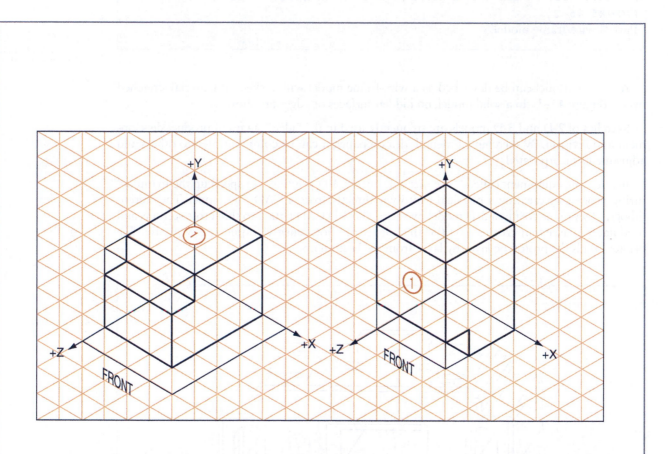

FIGURE 45–4
View coordinates.

With World Coordinates, the second system, the axes of the object remain attached to the same view even if the object is rotated (Figure 45–5).

World Views are very useful when programming CNC machines. Axis designations are necessary for electronic systems to assign movement to machine elements. For example, the depth axis (z) will be horizontal on a profile milling machine, while it will be vertical on a drilling machine.

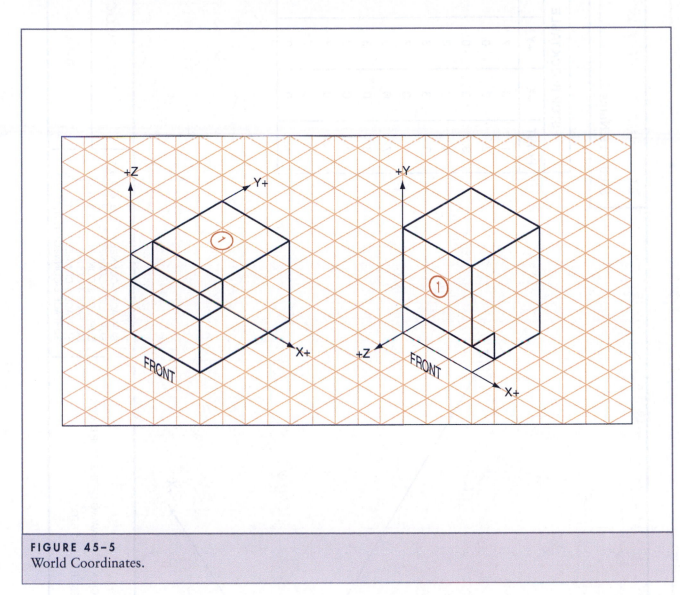

FIGURE 45–5
World Coordinates.

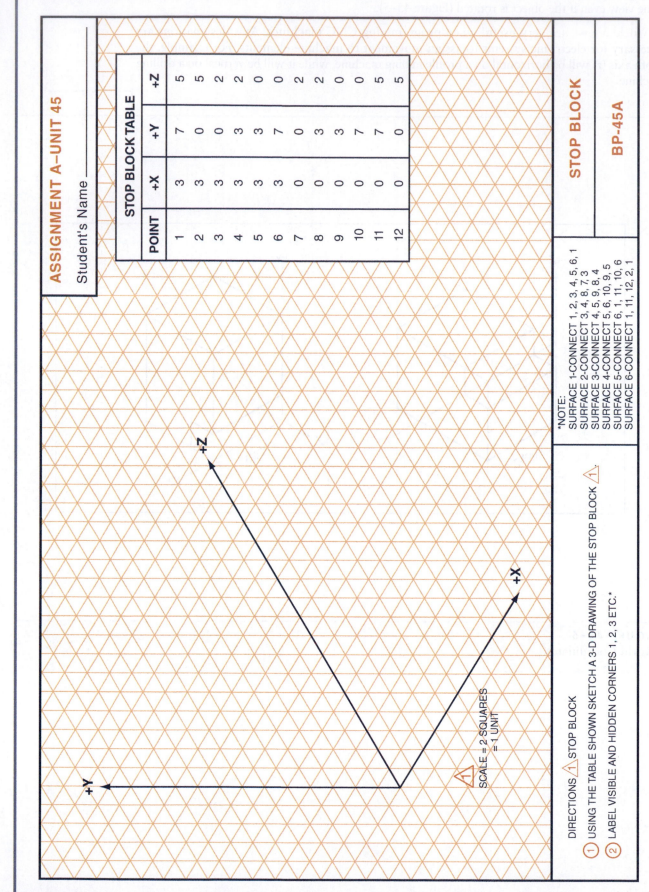

ASSIGNMENT A–UNIT 45

Student's Name _____

STOP BLOCK TABLE

POINT	+X	+Y	+Z
1	3	7	5
2	3	0	5
3	3	0	2
4	3	3	2
5	3	3	0
6	3	7	0
7	0	0	2
8	0	3	2
9	0	3	0
10	0	7	0
11	0	7	5
12	0	0	5

STOP BLOCK

BP-45A

DIRECTIONS △ STOP BLOCK

① USING THE TABLE SHOWN SKETCH A 3-D DRAWING OF THE STOP BLOCK △.

② LABEL VISIBLE AND HIDDEN CORNERS 1, 2, 3 ETC.

SCALE = 2 SQUARES = 1 UNIT

*NOTE:
SURFACE 1-CONNECT 1, 2, 3, 4, 5, 6, 1
SURFACE 2-CONNECT 3, 4, 8, 7, 3
SURFACE 3-CONNECT 4, 5, 9, 8, 4
SURFACE 4-CONNECT 5, 6, 10, 9, 5
SURFACE 5-CONNECT 6, 1, 11, 10, 6
SURFACE 6-CONNECT 1, 11, 12, 2, 1

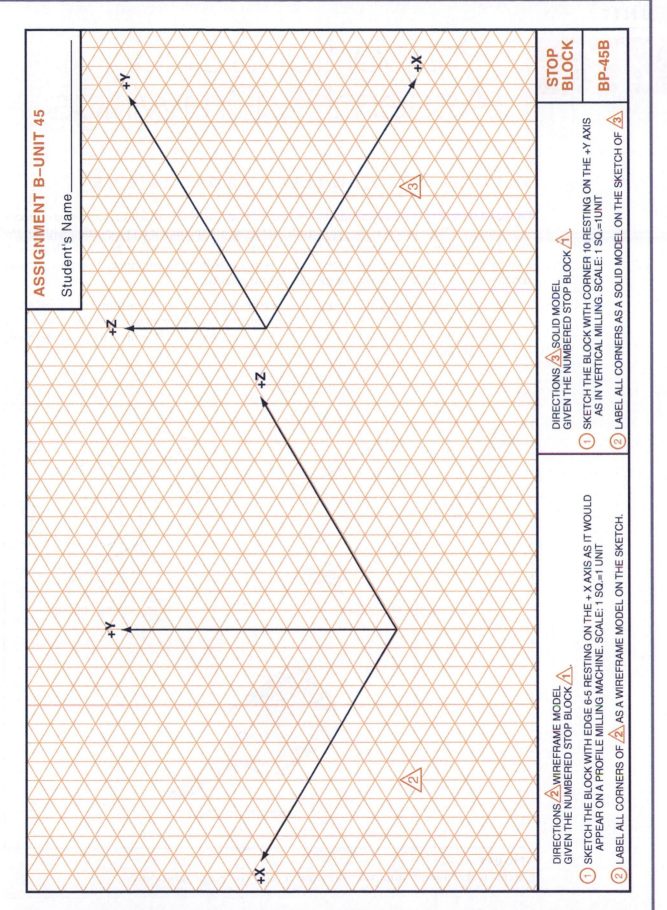

ASSIGNMENT B—UNIT 45

Student's Name

STOP BLOCK

BP-45B

DIRECTIONS △3 SOLID MODEL
GIVEN THE NUMBERED STOP BLOCK △1.

① SKETCH THE BLOCK WITH CORNER 10 RESTING ON THE +Y AXIS
AS IN VERTICAL MILLING. SCALE: 1 SQ.=1 UNIT

② LABEL ALL CORNERS AS A SOLID MODEL ON THE SKETCH OF △3

DIRECTIONS △2 WIREFRAME MODEL
GIVEN THE NUMBERED STOP BLOCK △1.

① SKETCH THE BLOCK WITH EDGE 6-5 RESTING ON THE + X AXIS AS IT WOULD
APPEAR ON A PROFILE MILLING MACHINE. SCALE: 1 SQ.=1 UNIT

② LABEL ALL CORNERS OF △2 AS A WIREFRAME MODEL ON THE SKETCH.

UNIT 45 | TWO-DIMENSIONAL AND THREE-DIMENSIONAL CAD SKETCHING

46

Proportions and Assembly Drawings

SKETCHING IN CORRECT PROPORTIONS

Freehand sketching requires judgment about how the sizes of different features are to be correctly proportioned. With practice, the eye can adequately create and maintain proportion without the use of superfluous, time-consuming devices. Usually, an approximation by eye is adequate. If a grid paper is used, each block generally represents a definite measurement. Still more precise sketches may be made by the use of measuring instruments.

One simple technique of sketching a part in correct proportion is illustrated in Figure 46–1. In this case, the handle diameter is equivalent to three units for height and depth and four units for length (width). The metal ferrule is proportionally 1 1/4 units wide, and the blade is 1 3/4 units long.

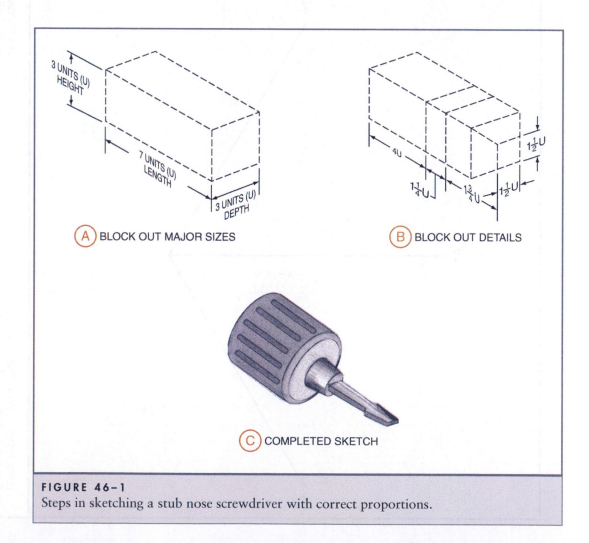

(A) BLOCK OUT MAJOR SIZES

(B) BLOCK OUT DETAILS

(C) COMPLETED SKETCH

FIGURE 46–1
Steps in sketching a stub nose screwdriver with correct proportions.

The three major steps in drawing this short blade screwdriver with accurate proportions include:

- Blocking out the major sizes
- Laying out the units for each detail
- Roughing out and then completing the sketch

PICTORIAL ASSEMBLY DRAWINGS

Most mechanisms and structures are composed of several individual parts that are connected together into subassembly and assembly drawings (see Unit 33). Examples of three types of assembly drawings follow.

The piston pin system shown in Figure 46–2 is an example of a sketched pictorial assembly drawing.

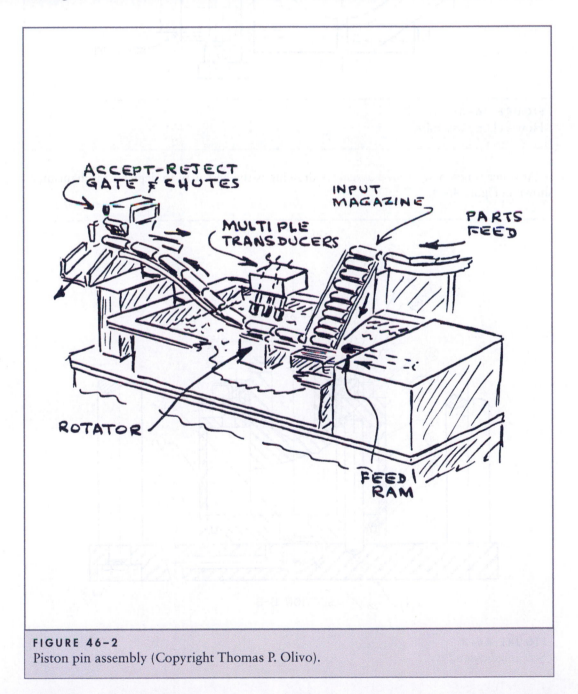

FIGURE 46–2
Piston pin assembly (Copyright Thomas P. Olivo).

The housing clamp **assembly** drawing shows a sketched top view (Figure 46–3).

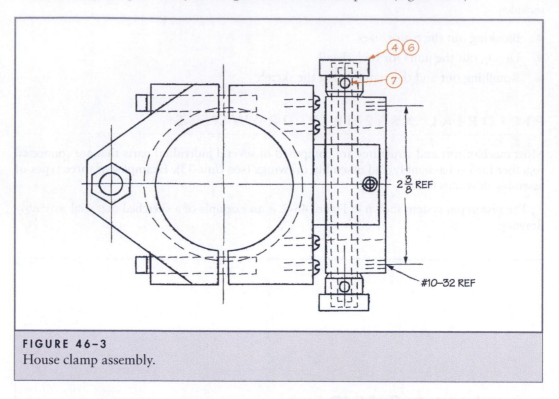

FIGURE 46-3
House clamp assembly.

At other times, a **sectioned assembly** drawing is the best way to depict a mechanism, as shown in Figure 46–4.

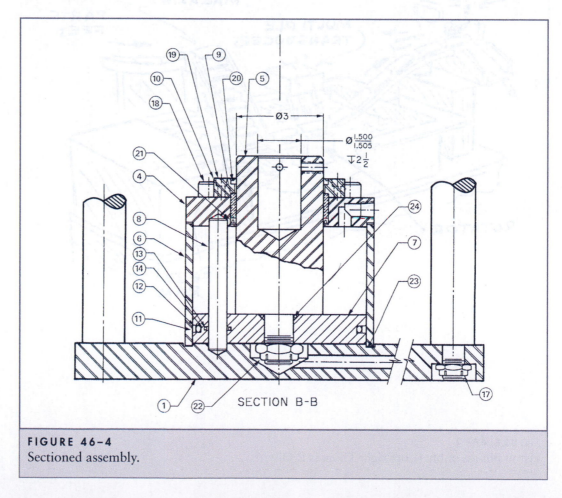

FIGURE 46-4
Sectioned assembly.

Exploded assembly drawings such as that in Figure 46–5, are used to show the order in which parts are to be put together. Note that this type of drawing is usually shaded.

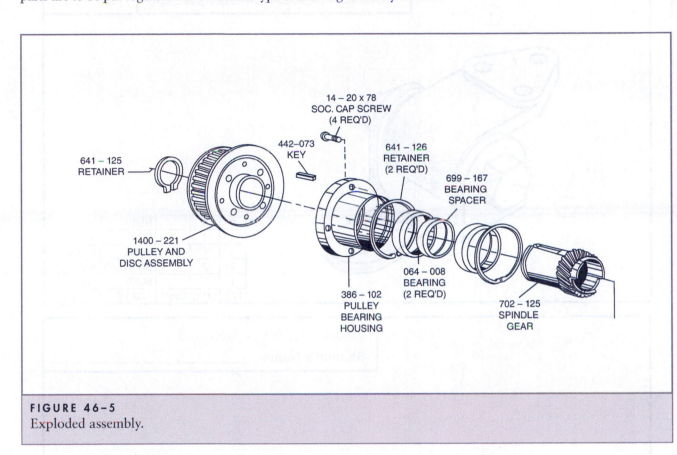

FIGURE 46–5
Exploded assembly.

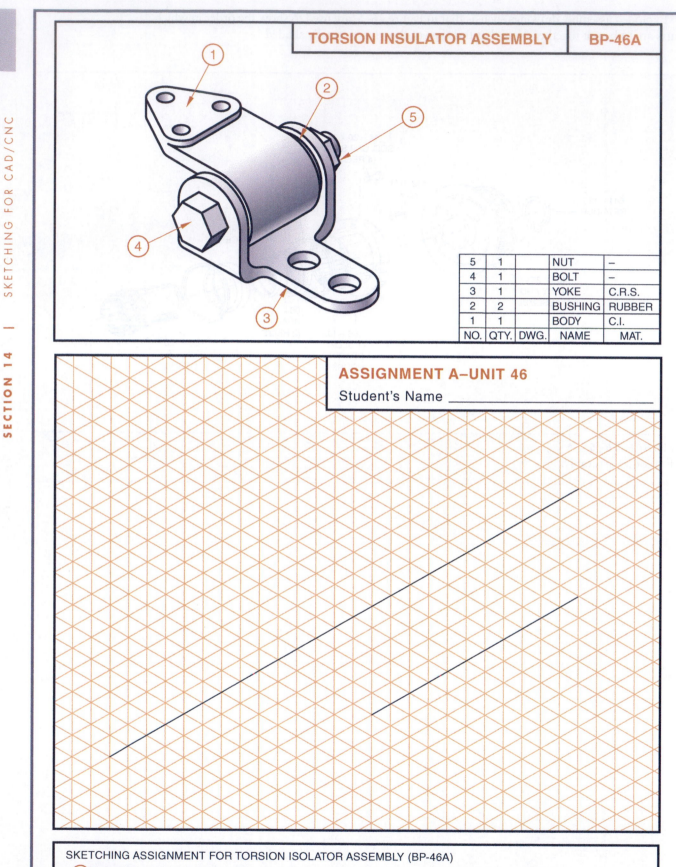

| | | TORSION INSULATOR ASSEMBLY | BP-46A |

NO.	QTY.	DWG.	NAME	MAT.
5	1		NUT	–
4	1		BOLT	–
3	1		YOKE	C.R.S.
2	2		BUSHING	RUBBER
1	1		BODY	C.I.

ASSIGNMENT A–UNIT 46

Student's Name _____

SKETCHING ASSIGNMENT FOR TORSION ISOLATOR ASSEMBLY (BP-46A)

1 SKETCH AN EXPLODED ISOMETRIC OF **BP-46A**. START WITH DARKENED GRID LINES.

2 LABEL EACH COMPONENT USING A LEADER LINE.

ASSIGNMENT B–UNIT 46

Student's Name _____

PLUNGER
HOUSING
SUBASSEMBLY

BP-46B

SKETCHING ASSIGNMENT FOR BP-46B

① FINISH THE EXPLODED ISOMETRIC DRAWING OF ALL THE PARTS THAT ARE IN THE BLOCKED-IN AREA.

② DRAW THE LEADERS, LETTER THE NUMBER OF EACH ONE OF THE PARTS, AND ADD SHADING TO THE SKETCH.

UNIT 46 | PROPORTIONS AND ASSEMBLY DRAWINGS

Glossary

A

Accuracy. A number of measurements that conform to a specific standard.

Aligned Dimensioning. A method of dimensioning a drawing. All dimensions are placed to read from the bottom or right side of the sheet.

Allowance. A prescribed, intentional difference in the dimensions of mating parts. *Positive allowance* represents the minimum clearance between two mating parts. *Negative allowance* represents the maximum interference between parts.

Alphabet of Lines. A universally accepted standard set of lines used alone or in combination to make a technical drawing.

ANSI (American National Standards Institute). A major professionally recognized body in the United States that establishes industrial standards. Represents the United States in ISO standards setting.

ASME. American Society of Mechanical Engineers publishes ANSI standards and establishes its own engineering standards.

Assembly Drawing. A drawing containing all the parts of a structure or machine when it is assembled. Each part is drawn in the relative position in which it functions.

Auxiliary View. The true projection of a surface that does not lie in either a frontal, horizontal, or profile plane. A primary auxiliary view lies in a plane that is parallel to the inclined surface from which features are projected in true size and shape.

Axis. An imaginary line that relates to particular features around which they rotate.

B

Base-Line Dimensioning. A system of dimensioning the features of a part from a common set of datums.

Basic Dimension. A theoretically exact dimension that is used to describe the size, shape, or location of a feature.

Basic Size. The exact theoretical size from which all limiting allowances and tolerances are made.

Bend Allowance. The amount of material required to form a sheet metal part to a specific radius.

Blueprint. Any copy of an original drawing, including hard copies.

C

CADD. Computer-Aided Drafting and Design.

CAM. Computer-Aided Manufacturing.

Chain Dimensioning. Successive dimensions that extend from one feature to another. Tolerances accumulate unless a note indicates "Tolerances are not cumulative." CNC using incremental dimensioning does not accumulate tolerance, as the CPU compensates for machining errors.

Cartesian Coordinates. A 2-D system of locating points using "X" and "Y."

CIM. Computer-Integrated Manufacturing combining CADD/CAM and robotics.

CNC. Computer-Numerical Controlled machine processing.

Computer. An electronic device capable of making logical decisions according to programmed control input.

Computer-Aided Drawing (CAD). The interlocking of functions, products, and systems of individual computers (software and hardware) into hierarchical order applications. Control of design, processes, and products by a hierarchical order of computer input.

Computer-Aided (Complete) Manufacturing (CAM) Application of CNC and CADD to all facets of manufacturing inclusive of and ranging from product/process design, production, inspection and assembly, marketing, etc., to plant management.

Constructed Parts. Usually refer to components made for use in large scale, one-of-a-kind structures made from reinforced steel, concrete, wood, masonry, and composite materials.

Conversion Tables. Cross-referenced equivalent mathematical values in two or more systems of measurement. Tables used to convert from U.S. customary units to metric values and vice versa.

Countersunk Hole (CSK). A hole that is machined with a cone-shaped opening to a specified outside diameter and depth. A conical recess that provides an angular seat for flathead screws and other cone-shaped objects.

CPU (Central Processing Unit). Center of the computer where all control functions and calculations take place.

Cutting Plane Line. A thick, broken line representing an imaginary plane cutting through an object. The resulting view is a section view.

D

Datum. An exact reference point from which a line, plane, or feature dimension is taken or a geometric relationship is established.

Decimal-Inch System of Dimensioning. A system in which dimensions are given, where subdivisions of the inch or millimeter are in multiples of ten; such as 0.1″, 0.01″, and 0.001″.

Detail Drawing. A drawing that includes all views needed to accurately describe a part and its features. Dimensions, notes, materials, heat treatment, quantity, etc., are included.

Dimension. Size and location measurements that appear on a drawing.

Dimensional Measurements. Measurements relating to plane surfaces; multiple surface objects; straight and curved lines; areas; angles; and volumes.

Drafters. Translate data graphical ideas and concepts of engineers, technicians, designers, and architects into detailed manufacturing or construction drawings.

Dual Dimensioning. A technique of giving all dimensions on a drawing (using soft conversion dimensions) in both customary inch and SI metric units of measurement.

E

Engineering Work Order. Instructions that give permanent change information from an original drawing. A drawing change may be called a revision, change notice, or alteration. Information is recorded in the change block area of the title block.

Exploded Drawing. A pictorial drawing showing each part of an assembled unit in its location in the order of assembly.

F

Fabrication. Layout and assembly of parts of a structure away from a job site.

Feature. The geometric form of a portion of a part that identifies particular characteristics at that location.

Feature Control Symbol. A simplified form of a geometric characteristic(s) combined with tolerance information and datum references that appear on blueprints in an enclosed rectangular box.

First-Angle Projection. The (SI [ISO]) metric standard of projection from quadrant I. Each view is drawn as projected through the object.

First-Angle Projection Symbol. The standard SI symbol ($\ominus\!\!-\!\!\oplus$) indicating that the drawing is made by the metric first-angle projection method as opposed to third-angle projection.

Fit. The relationship between two mating parts produced by interference, clearance, or transition.

Fixture. A device used to accurately position and to hold a piece part of subsequent machining processes.

Flange. An edge, rim, or collar usually at a right angle to the body of the part.

Forging. The forming of heated metals by hammering between forging dies (or rolling).

Form Tolerancing. The permitted variation of a part feature from the exact form identified on the drawing.

Fusion Welding. The joining at a particular location of metals that are brought up to fusion temperatures.

G

Geometric Control Symbols. ANSI geometric characteristics symbols grouped according to form tolerance, location, tolerance, and runout tolerance.

Geometric Tolerancing. Specified tolerances that apply shape, form, or position of specific features of a part.

H

Hard Copy (Blueprint). An end-use copy of an original drawing.

Hard Conversion. Features of an object designed to conform to standard sizes in the other system of measurement (SI metric or customary inch). Objects designed by hard conversion are not interchangeable between two measurement systems.

I

Interchangeable Parts. Allows parts or mechanisms made anywhere in the world to be exchanged with another part made to the same specifications.

Intersections. The lines created at the point where two or more objects (such as two planes) meet or pass through each.

ISO (International Organization for Standardization). Composed of about 100 member nations, including the United States, developing metric standards.

Isometric Drawing. A pictorial drawing of an object with all three axes making equal (120 degrees) angles with the picture plane. Measurements on each axis are made to the same scale.

J

Jig. A work-holding and tool-positioning and guiding device.

K

Keyslot. A slot machined in a shaft to receive a Woodruff (semi-circular type) key.

Keyway. A groove cut into the hub area of a mating part to receive that portion of a key that extends beyond the shaft.

L

LAN (Local Area Network). Either a hard-wired or wireless network connecting a number of computer workstations to a server.

Laser. Acronym for light amplification by stimulated emission of radiation. One of the most important tools of the micro-technology age. It makes extremely fine accurate cuts. A device for producing light by emission of energy stored in a molecule or atomic system when stimulated by an input signal.

Lay. The direction of the dominant surface pattern of the cutting or forming tool in the manufacture of a part.

Limits. The maximum and minimum or extreme permissible sizes of acceptance of a part.

Location Dimension. A dimension that identifies the position of a surface, line, or feature in relation to other features.

M

Machining Center. A group of high-precision, multifunctional machines for performing specific operations without needing secondary operations.

Manufactured Parts. May involve casting, forging, and/or machine tools to produce objects from metallic, plastic, or organic materials.

Mass Production. The manufacturing, construction, or fabrication of large quantities of identical parts or assemblies in a short time span.

Maximum Material Condition (MMC). A condition where a part feature contains the maximum amount of material; for example, minimum hole diameter and maximum diameter of the mating.

Measured Point. The specific point at which each measurement ends.

Metrication. The international movement toward the universal adoption of the ISO metric system.

MIG-Welding. Acronym for Metal Inert Gas welding, similar to TIG welding. The electrode is a filler wire that is fed into a weld.

Milling. A material-removal process requiring a rotating cutting tool. Machining process performed on the milling machine.

Multiview Projection. The projection of two or more views of an object upon the picture plane in orthographic projection.

N

Nominal Size. A size that represents a particular dimension or unit of length without specific limits of accuracy.

Notes. A supplemental message found on a drawing that provides technical information. *General notes* relate to the entire drawing. *Local notes* give processing information relating to a specific area of the object.

Numerical Control System Functions. Basic NC system functions include controlling machine and tool positions, speeds and feeds, shutdown, and recycling; establishing operation sequences; monitoring tool performance; and producing readouts.

O

Oblique Drawing. A pictorial drawing of an object that has one of its principal faces parallel to the plane of projection. Features on this face are seen and projected in true size and shape. The features on other planes are drawn oblique to the plane of projection at a selected angle.

Ordinate Dimensioning. An arrowless rectangular datum dimensioning system. Dimensions originate from two or three mutually related datums or zero coordinates.

Origin. The zero point in the Cartesian coordinate system.

Orthographic Projection. The standardized method of representing an object with details by projecting straight lines perpendicular from the object to two or more planes. Third-angle orthographic drawings are standard in the United States and Canada.

P

Part Programming. Producing complete input information to machine a part by CNC.

Perspective Drawing. A pictorial drawing on which the features of an object appear to converge in the distance toward one, two, or three vanishing points.

Pictorial Drawing. A one-view drawing showing two or more faces of an object. An easily understood drawing that provides a quick visual image of an object. The three major types of pictorial drawings are *axonometric*, *oblique*, and *perspective*.

Plane of Projection. An imaginary plane placed between the observer and the object. Assumed to be perpendicular to the line of sight.

Positional Tolerancing. The extent to which a feature is permitted to vary from the true or exact position indicated on the drawing.

Print. A copy of an original technical drawing. Generally called a *blueprint* regardless of the type or reproduction process.

Profile. The shape or contour of a surface. The profile is viewed on a plane that is perpendicular to the machined surface.

Program (Computer). Step-by-step instructions to a computer for solving a problem by using input information from manual operation or as contained in computer memory.

Projection (Construction) Lines. Very lightly drawn non-reproducing layout lines used in instrument drawings and sketching.

R

Reaming. A machining process for finishing a hole to close dimensional and geometrical tolerancing standards.

Rectangular (Cartesian) Coordinate System. A four-quadrant (x, y, and z axis) system with vertical and horizontal planes. A system of representing plus and minus values to control NC machine tools and processes.

Reference Dimension (REF). Dimension given on a drawing to provide general information. Not used in the manufacture of a part.

Regardless of Feature Size (RFS). A condition of conforming to the tolerance of position or form regardless of the feature size within its tolerance.

Resistance Welding. A metal-joining process that depends on the resistance of a metal to the flow of electricity to produce the heat for fusing the metal parts.

Rib. A design feature (reinforcing member) to strengthen a fabricated or molded part.

Robotics. A programmable, multifunctional manipulator designed to move the materials, parts, tools, or specialized devices through motions performing a variety of tasks.

Roughness. The fine surface irregularities produced by cutting tools, feed marks, or other production processes.

S

Scale. A device (mechanical or computer generated) used to draw objects smaller or larger than their actual size.

Sectional View. A view produced by passing an imaginary cutting plane into an object at a selected location, removing the cut-away section, and exposing the internal details.

Shading. Additional lines, patterns, or tonal graduation to heighten illusion of depth.

SI Metric System. The internationally accepted decimal system of measurements built on seven basic units of measurement. The SI metric system has been adopted by the ISO as the basis for its drafting standards.

Soft Conversion. Direct conversion of a measurement in one system to its equivalent in the other system. Provides equivalent measurements in two systems without regard to the standardization of parts in both systems.

Solid Molding. A refined type of computer modeling that recognizes both exterior and interior features of an object.

Specifications. Stated requirements (standards) that must be satisfied by a product, material, process, structure, component, or system.

Spotfacing. Producing a flat bearing surface over a small area of a part for purposes of seating a bolt or shoulder of a shaft.

Stretchout. A flat pattern development that, when formed, produces a sheet material product of a particular size and shape.

Surface Texture. The designation of the characteristics and conditions of a surface. Deviations from a nominal profile that forms the pattern of a surface. Deviations produced by *roughness, waviness, lay,* and *flaws.*

Symbol. A letter, numeral, character, or geometric design that is accepted as a standard to represent a feature, operation, part, component, etc.

Symmetrical (SYM). A condition where features on both sides of a center line are of equal form, dimensional size, and surface quality.

T

Tabular Dimensioning. An ordinate dimensioning system in which all dimensions are taken from datums. Each dimension on a drawing is identified by a reference letter with a corresponding numerical value found in a table or chart.

Taper. The uniform change in size along the length of a cylindrical part, as in the case of a taper shank, or on a flat surface, such as a wedge-shaped key.

Tapping. Forming a uniform internal thread using a hand or machine tap.

Technical or Engineering Drawing. Furnishes a graphical and mathematical description of the size, shape, and form of an object as well as other essential information needed to produce a physical object to the degree of precision specified.

Technical Sketch. A quick and inexpensive way to convey graphical instructions.

Technician. A person who works in a technical field as a member of an architectural or engineering support team.

Thread Class Number. A number appearing on a drawing with a thread note to indicate the required degree of accuracy between mating features of an external and internal thread.

Thread Form. The shape or profile of a thread. The most common thread form is based on a 60-degree included angle. American National Standard Unified threads and ISO metric threads use this angular form.

TIG-Welding. Acronym for Tungsten Inert Gas welding. A gas-shielded arc-welding process.

Title Block. An area on a drawing containing details and specifications complementing the information given on the drawing. An information block usually located at the bottom right corner of the drawing.

Tolerance. A total permissible variation in the size of a part. The dimensional range within which a feature or part will perform a required function. The difference between upper and lower limits of a dimension.

True Position. The theoretically exact reference location of a feature.

Typical (TYP). A dimensional feature that applies to the locations where the dimension or feature is identical, unless otherwise noted.

U

Undercut. A recess at a point where a shaft changes size and mating parts fit flush against a shoulder.

Undimensioned Drawings. A technique used to produce parts and components that have a continuously changing profile, where the use of conventional dimensioning is impractical. Undimensioned parts (particularly in the aerospace industry) are often CNC-machined by controlling input information to the machine control unit (MCU).

Unidirectional Dimensions. Dimensions placed to read in one direction.

Unidirectional Tolerance. A single-direction tolerance that is specified as above (+) or below (−) a basic size.

USB Flash Drive. A miniature portable device similar in use to a computer hard drive. [Securely stores computer data magnetically.]

V

Visualization. The process of creating a 3-D image of an object given its multiple views or imaging multiple views of an object given its 3-D drawing.

W

Waviness. The widest-spaced component of the surface; it covers a greater distance than the roughness-width cutoff.

Welding Symbol. A standard designation consisting of a reference line, arrow, and tail. Other symbols, code letters, dimensions, and other notes are added.

Wire Frame Model. A refined type of computer modeling. The computer recognizes and produces only exterior surfaces.

Working Drawing. A multiview orthographic drawing that provides details for the design, manufacture, construction, or assembly of a part, machine, or structure.

Z

Zero Point. The defined origin point of a CNC machine.

Zip Drives. A removable magnetic device that securely stores computer data. Zip drives are portable and easy to use.

Zoning. A system of drawing sheet perimeter numbering and lettering establishing locations of objects.

Index